PAITA,
OUTPOST *of* EMPIRE

PAITA,
OUTPOST *of* EMPIRE

*The Impact of the New England Whaling Fleet on the
Socioeconomic Development of Northern Peru, 1832-1865*

William L. Lofstrom

MYSTIC SEAPORT MUSEUM, INC. MYSTIC, CONNECTICUT 1996

Mystic Seaport Museum, Inc.
75 Greenmanville Ave., P.O. Box 6000
Mystic, CT 06355-0990

Cataloging in Publication data

Lofstrom, William L.
    Paita, outpost of empire : the impact of the New England
whaling fleet on the socioeconomic development of northern Peru,
1832-1865 / William L. Lofstrom. – 1st ed. – Mystic, Conn. :
Mystic Seaport Museum, 1996.
        p. : ill., maps ;   cm.
    Bibliography: p.
    Includes index.

    1. Paita (Peru) - Social conditions - 19th century. 2. Paita
(Peru) - Economic conditions - 19th century. 3. Whaling - New
England - History. 4. U.S. Consulate. Paita, Peru. I. Title.

HN350.P3L6
ISBN 0-913372-74-9

Contents

Maps, Figures, and Plates follow page 64

Foreword

Much United States maritime history seems to regard South America principally as an obstacle to easy access to the Pacific Ocean. So, indeed, the Peruvian port of Paita, 13,000 miles by sea from New England, might seem an unlikely subject for Mystic Seaport. How could a small port on the arid northern coast of Peru be of significance in the maritime history of the United States? Certainly Paita does not belong on the list with Liverpool, or Le Havre, or Rio de Janeiro, or Callao, or Canton as an entrepôt for trade with the United States in the nineteenth century.

Yet, Paita was an important place for a characteristic New England maritime industry, whaling. When U.S. whaleships first entered the Pacific Ocean in the 1790s and began to extend the nation's influence throughout the South Pacific, they sought sources of provisions among the islands and settlements in the region. After Peruvian independence, the former Spanish port at Paita welcomed whaleships, providing them with potatoes and other foodstuffs, serving the liberty crews in its bars and brothels, and sometimes offering repair facilities for disabled ships.

By the 1830s the port had enough American activity to warrant establishment of a U.S. consulate to address diplomatic and commercial concerns and to protect the rights of U.S. seamen. From the 1830s to the mid-1860s more than 300 U.S. whaleships called at Paita for a total of nearly 1,200 visits. Much of the data from which we can reconstruct the life of the port comes from the consular documents, which report the vessels that called and the occasions when the consul assisted Americans in Paita.

The five men who served as U.S. consul in Paita

FOREWORD

between 1839 and 1874 are the major characters in this study, not simply for their efforts to represent U.S. legal and commercial interests there. As expatriates, they characterized some of the motivations that lured foot-loose Americans to all corners of the world, whether it was the commercial prosperity that attracted an Alexander Ruden, Jr., or the scientific and medical opportunities sought by a Dr. Charles F. Winslow, or even a patronage position such as that accepted by Civil War veteran John Murphy. In coming to know these unsung individuals even a little, we can begin to understand both how the consular mission operated overseas and what drew men to a new life in distant foreign realms.

With a long career in the U.S. foreign service in several South American nations, as well as a clear understanding of the economic history of South America, William Lofstrom brings ideal credentials to this study. On the one hand, his sensitive analysis of the changing economic fortunes of the port helps us recognize what has been, in reality, an enduring and significant relationship between the Northern and Southern Hemispheres. At the same time, Lofstrom's work is especially valuable in applying Carlos Assadourian's concept of the *espacio económico* to a study of economic and social history in the English language. U.S. maritime historians often use the entrepôt concept to describe the workings of U.S. ports, and Assadourian's construct ties the port into its hinterland in an organic way. Just as the port of Paita stands out in high relief against the arid *tablazo* of the Peruvian coast, William Lofstrom has clearly reconstructed the intercultural social and economic network of Paita in this model monograph.

ANDREW W. GERMAN
MYSTIC SEAPORT MUSEUM

Preface

In a pilgrimage to the venerable George Peabody Library in Baltimore in 1990 to search for nineteenth-century graphic material to illustrate my book on *Cobija y el litoral boliviano vistos por ojos extranjeros, 1825-1880* (La Paz, Fundacion Cultural Quipus, 1991), I happened across the five beautiful lithographs depicting the Peruvian port of Paita in 1837 that ilustrate this volume. I was struck by the fact that the members of a French scientific expedition should find Paita to be of sufficient interest to dedicate several pages of narrative and five folio pages of their historical atlas to it, when today it is of so little economic importance. I had never been to Paita. No one that I know who has been there found it in any way remarkable. Until that moment in Baltimore my only awareness of the northern port was a vague recollection of a grandiose fishing project based in Paita that the populist military government of General Juan Velasco Alvarado had proposed in the 1970s, when I was serving in the American Embassy in Lima.

My curiosity aroused, I sought an explanation for Paita's former importance in the American consular dispatches generated in that port between 1833 and 1874. The fact that the United States maintained a consulate in Paita was, in itself, suggestive of Paita's former significance. From the consular reports I learned of Paita's pivotal role as a port of call for the numerous New England whalers that ranged the Pacific prior to the American Civil War, in search of the elusive sperm whale. The dispatches reveal the dimension of Paita's role as a service economy catering

PREFACE

to foreign vessels. They also reveal the presence of a surprising number of foreigners, particularly Americans, in northern Peru in the early decades of the nineteenth century. The consular reports attest to the existence in Paita during this period of an interesting expatriate institution, the American Consular Hospital. Finally, they suggest that the American presence in Paita, beginning in the 1830s, was a catalyst in the subsequent modernization of northern Peru and in its integration into the export-oriented economy that characterized the second half of the nineteenth century.

The eminent Peruvian historian and bibliographer Jorge Basadre defined true economic history as the "theory, sociology, geography, anthropology and demography of the past."[1] This apt description has been my guide in the preparation of this study of the role played by the Peruvian port of Paita and its hinterland, in the nineteenth century, as a socio-economic bridge between Peru and the United States.

The classical characterization of different regions of Spanish America as either core or pheriphery can be useful as an analytical or didactic tool, especially when dealing with the colonial period and the nineteenth century. While generally used to differentiate between regions where economic and political power were concentrated, like Mexico and Peru, and isolated appendices of empire like Chile and the Californias, the core-periphery analysis seems equally valid when applied to a more limited geographical area such as the Viceroyalty of Peru in the late-eighteenth century, or the republic that was created within that jurisdiction in 1821.

Within this more limited scope, it becomes clear that virtually every area outside the Lima-Callao metropolis must be considered peripheral, even though cities such as Arequipa and Cuzco played important roles in regional economic and political developments. The appealing simplicity of the core-periphery dichotomy should not, however, obscure the fact that in Peru some peripheral areas were very different from others. This difference was in part a function of the physical space that separated them from Lima, the looseness of their economic and political links with the metropolis, whether or not they were situated on a frontier, and the nature of their ties to neighboring countries. In addition, with the opening up of the former Spanish empire to foreigners, some areas that were clearly peripheral in terms of their relationship with the capital began to develop significant economic and social ties with a far broader world, a phenomenon that in some cases served to further distance and differentiate them from their political and economic core.

The northern port of Paita and its hinterland is one such region. It is a textbook example of a peripheral area, an outpost of empire, whose sustained economic and social ties with foreigners set it apart from the rest of Peru, stimulated important socio-economic changes, and had a significant impact on the country as a whole. In the early nineteenth century Paita was a long way from Lima, physically, politically, and socially. It was largely unaffected by the chaos of the wars of liberation and the *caudillismo* of the early republic. It shared little in the benefits of the guano boom (1840-56) and was virtually untouched by the War of the Pacific (1879-83) and the Chilean occupation of Lima.

Instead, Paita and its *espacio económico* - to borrow a useful concept from the Argentine historian Carlos Sempat Assadourian - quietly attended to business. For a third of a century business as usual for Paita was the care and feeding of the men and ships of the New England whaling fleet. In the process, Paita attracted a community of foreign residents, drawn by the lure of money to be made from the whaling trade and ancillary activities. They and their allies among the local elite accumulated capital - modest sums in contrast to the fortunes made throughout Peru's boom and bust economic history - but adequate to finance modern port facilities, improved communications, urban amenities, and a gradually expanding agricultural export sector.

Assadourian's thoughtful analysis of the economic history of colonial Peru serves as a theoretical framework for understanding Paita's role as a socio-economic bridge between Peru and the United States in the nineteenth century.[2] His concept of *espacio económico* - in contrast to the more traditional concept of political or national space - is applied to the entire viceroyalty, but is equally applicable to a smaller area like the Paita hinterland. Paita's *espacio económico* - which embraced the Intendency (later Department) of Piura as well as the Andes of northern Peru and southern Ecuador - was characterized by a well-integrated, largely self-sufficient internal market of great magnitude, which was driven by the significant need of an "economic pole" (the port of Paita and the capital, Piura) for complementary products (provisions, water, and marine stores). This economic hinterland was impacted by what Assadourian calls *efectos de arrastre* - side effects - that in time resulted in a process of capital accumulation that served as a true point of departure for economic growth.

Within this conceptualization, the New England whalers helped to transform Paita and northern Peru. The economic and social impact of their sustained presence there converted an isolated, peripheral port that had been an outpost of the Spanish empire into an outpost of the incipient North Atlantic economic empire, whose full realization was to come later in the century.

The challenge that impelled the research for this study comes from Basadre's dictum that "historiography that does not go beyond the description of facts and particular situations may survive, perhaps, in the area of political history; but lacks a future in the [area] of economic history."[3]

The interpretation of facts that Basadre demands is not easy, especially - as in the case of Paita and its *espacio económico* - when the documental base for that interpretation is skewed and sketchy. The Peruvian government did not begin to publish economic data until 1891. Thus, one of the primary sources for information about the economic life of Paita in the early nineteenth century comes from the dispatches of United States consuls in that port between the years 1833 and 1874.

The consuls were primarily interested in the whaling industry and Paita's role as a port of call. Complementary information on the economy of Paita, the *espacio económico* of northern Peru, and the impact of the foreign presence there is also found in consular dispatches and in travel accounts, but it is often impressionistic and sometimes unreliable. In the final analysis, perhaps Professor Richard Saeger of Lehigh University is correct in his dictum that often "the most important research tool is historical imagination."[4] The fact that the history of nineteenth-century Peru is still the least exploited of any of the major national histories of the republican era - "a dark age within the Latin American dark ages" in the words of Paul Gootenberg - makes the case for telling the story of Paita even more compelling.[5]

This study is organized along two parallel but integrated tracks or story lines. The first track - comprising chapters one through three, five, six, eight, and ten - describes Paita as a entrepot and how the New England whaling industry altered the port from 1832 to 1865, stimulating its development as a service economy. This track also relates how, beginning in about 1860, Paita was gradually transformed from a service economy to an export economy focused on the provision of goods for European and North American markets.

The second track, comprising chapters four, six, seven, nine, and eleven, deals with the five individuals who served as American consul in Paita during this period, recreating as well as possible their family backgrounds, economic interests, professional training, and performance as consul in Paita. These profiles of expatriate Americans and their varied experiences in Peru are complemented by extensive data in chapter seven on other foreigners, primarily from the North Atlantic world, who were drawn to Paita by many forces, primarily the lure of economic opportunity. The conclusion, chapter twelve, attempts to bring these two story lines together. The picture that emerges confirms the importance of the New England whaling fleet on the socio-economic development of northern Peru, and Paita's role as a socio-economic bridge between Peru and the United States in the nineteenth century.

I owe the following people a debt of gratitude for their inspiration and help in the preparation of this study. Tom Orum's stories of the involvement of Yankee traders in the Brazilian Amazon in the early nineteenth century inspired me. His invaluable guidance in the art of genealogical research and the many resources available in Washington helped make this story possible. Betsy Franks and Greta Wilson of the Department of State Library, Washington, D. C., were extremely helpful in securing material through inter-library loan and in answering reference questions. Judith Downey, Curator of Manuscripts, and the staff of the New Bedford Whaling Museum in New Bedford, Massachusetts, generously made available to me the Museum's impressive manuscript collection of ships papers and other resources. Steve Oltmann prepared the maps that accompany this volume. Finally, Erick Langer, Susana Aldana, and Christopher Paddack read early drafts and made many suggestions for significant changes that, I believe, have improved it considerably.

I am also indebted to the following institutions for their help in the course of my research, particularly the ferreting out of information on the unknown Americans whose presence in Paita during the first half of the nineteenth century contributed to building the socio-economic bridge between Peru and the United States: The Archive of the United States of America, Washington, D. C.; The Daughters of the American Revolution Library, Washington, D. C.; The Library of the American Institute of Architects, Washington, D. C.; The Historical Society of Washington, D. C.; The Columbus Memorial Library of the Organization of American States, Washington, D. C.; The Maryland Historical Society, Baltimore, Maryland; The George Peabody Library of the Johns Hopkins University, Baltimore, Maryland; the Maryland Historical Trust, Annapolis, Maryland; The Washington County Free Library, Hagerstown, Maryland; St. James School, St. James, Maryland; The New-York Historical Society, New York; The Buffalo and Erie County Historical Society, Buffalo, New York; The Kendall Whaling Museum, Sharon, Massachusetts; The Essex Institute and the Peabody Museum, Salem, Massachusetts; The Bostonian Society, Boston, Massachusetts; The University of Pennsylvania Archives and Records Center, Philadelphia, Pennsylvania; The Lloyd House, Alexandria, Virginia; The Special Collections Department of the University of Virginia Library, Charlottesville, Virginia; the Medical Archives of the New York University of Health Sciences, New York; the History of Medicine Section, Yale University Medical School, New Haven, Connecticut; the Ironbridge Gorge Museum, Ironbridge, Telford, Shropshire, United Kingdom; the Museum of American Textile History, North Andover, Massachusetts; and the Church of Jesus Christ of Latter Day Saints, Salt Lake City, Utah.

I am, of course, responsible for any errors of omission or commission related to the transcription and interpretation of the data provided by them.

1

Paita,
Outpost of Empire

As a settlement of Europeans, San Francisco de la Buenaventura de Paita is coeval with the Spanish conquest of Peru. The town's origins and its name, however, are indigenous. According to the pioneer nineteenth-century Peruvian geographer Mariano Paz Soldan, the word "paita" means desert, which would certainly be appropriate, but none of the other sources I have consulted suggests the derivation or meaning of the word. Obviously, however, the place "Paita" existed before the arrival of the Spanish.

The spelling of that name has varied over time. The variation "Payta" is frequently found in nineteenth-century North American maps, travel accounts and other documents. One traveler, in positing an explanation for this variant, hints at Paita's importance as a bridge between two cultures and two economies. According to this traveler, the "affluent" Spanish pronunciation of the first two vowels in Paita, as interpreted and transmitted to North Americans

CHAPTER 1

ears by the "impoverished" speech of New England whalers, resulted in the transcription of "Payta" on most United States maps of South America.[1]

The European Presence in Northern Peru

From the very dawn of the European presence on the Pacific coast of South America, Paita played an important role as a source of supplies for foreign vessels. William H. Prescott, in his *History of the Conquest of Peru*, relates that Francisco Pizarro, at the beginning of his famous second voyage of discovery off the coast of Peru in 1532, visited the bay and port of Paita. The Indians living there, upon learning of the presence of the Europeans, came out to Pizarro's ship in their *balsa* rafts with fruit, fish, and vegetables for the crew's mess.[2]

Pizarro's "discovery" of Paita had been preceded by a brief visit to Tumbes, on the southern coast of the Gulf of Guayaquil, where he had left a small contingent of men during his first voyage of discovery into the South Seas. Finding the colony at Tumbes destroyed, mistrustful of the Indians there, and reluctant to attempt a second settlement, Pizarro sailed on to the bay of Paita where, in the words of Sebastian Lorente, he found "a place that, because of its proximity to such an excellent port and its abundant resources, offered him a good base of operations." This "place" was probably the lower reaches of the Chira River, just north of Paita (see map 1).[3]

Despite its advantages, Paita was not chosen by Pizarro as his administrative headquarters in northern Peru. The port's absolute lack of water and the difficulty in defending it militarily probably influenced Pizarro to choose a well-watered inland site, which was to become the city of San Miguel de Piura. On his third expedition to Peru, Pizarro proceeded from Puerto Viejo in modern Ecuador to Puna Island in the Gulf of Guayaquil, and from there to the site of his first settlement at Tumbes. From

2

PAITA, OUTPOST OF EMPIRE

Tumbes the Spaniards continued 30 leagues (5,572 meters) south across the Talara peninsula to a site known as Tangarará - near present-day Sullana - on the south bank of the Chira River, where a town government was organized, land was allotted to Pizarro's men, and the local Indians were given in *encomienda* to the settlers. This site was later found to be malarial and subject to flooding, and in 1533 Piura was moved south to the banks of the Piura River, which eventually empties into the larger Bay of Sechura south of Paita. The second site of Piura was also abandoned in 1728 because of serious flooding associated with the El Niño phenomenon, and the town was moved downstream.[4]

Of the 40 or more oasis-like river valleys on the extremely arid coast of Peru, the Chira River valley that empties into the Bay of Paita is one of the most fertile. The Chira rises as the Piscobamba in the cordillera of the Andes of southern Ecuador, at the Páramo de Sabanilla, and after passing south of the city of Loja becomes the Catamayo. The Catamayo forms part of the border between Ecuador and Peru, and then, as the Chira, flows almost due west until it reaches the Bay of Paita. The lower Chira valley varies in width, from the rich agricultural bottom lands over four miles wide at Viviate and Monte Abierto to the narrow gorges at Amotape and Sullana (map 2).[5]

Of the numerous bays that line the northern coast of Peru, that of Paita is the best. The accounts of foreign travelers and adventurers all agree that the bay is large, secure, very accessible, deep, and that it has a good bottom (see map 3). A visitor in 1833 compared it to the bay at Valparaiso, but noted that it is larger. In 1854 the American consul described it as "beautiful, safe and commodious," noting that large vessels could lie at anchor within a few hundred yards of land. Another visitor in 1861 noted that his ship anchored in nine fathoms (54 feet or 18 meters) of water only one-half mile from shore, and an English botanist described it in 1864 as "the most secure and commodious anchorage of any port along the whole coast of Peru."[6]

The bay has a hard sand beach and is surrounded by a 150-foot high bluff, from which an arid, sandy plain - the *tablazo* - gradually rises to the foothills of the Andes. A constant wind blows sand and dust in all directions. This wind, and the dry, cool climate in winter (60 to 70 degrees) were thought to be conducive to pulmonary diseases in the nineteenth century, but in general the climate in Paita is healthy.[7]

The town of Paita is situated on the southern, protected side of the bay, at the foot of a promontory that for centuries has served as a landmark for mariners, the Silla de Paita. Throughout the late eighteenth and early nineteenth centuries the town's configuration remained basically the same (see map 4). It consisted of two or three unpaved streets running parallel to each other and to the shoreline, connected by irregular, narrow alleys (see plate 1). On the eastern edge of town, slightly separated from the other buildings, were the Convent of La Merced and the parish church of San Francisco de Paita. According to local legend, an image of the Virgin Mary in the Church of La Merced that had been slashed by English corsairs in 1741 still had, in 1812, an open bloody wound on the neck (see plate 2).[8]

Most of the 200 or so houses in Paita during this period were made of split bamboo over rough bamboo and wood frames, chinked and plastered with mud, a construction technique known as *bajareque*. One traveler described the town in the following fashion: "Imagine a series of the mud nests of the barn-swallow, set close together on a narrow, sandy beach, at the base of a ledge of pale-gray and disintegrating rock, with no sign of vegetation far or near, and you will get a pretty accurate picture of the town of Paita."[9]

PAITA, OUTPOST OF EMPIRE

The houses had steeply pitched, palm-thatched roofs, to accommodate the infrequent but often torrential rains associated with the El Niño phenomena in northern Peru, which in the nineteenth century were known as *años de aqua*. Many of the roofs were plastered with a gypsum made from firing the sedimentary selenite found in the region, which protected the thatch from the constant wind. The floors of many of the houses in Paita were plastered with the same material. Virtually the only two-story house during the late colonial period was that of the governor, facing the bay between the Convent of La Merced and the small fort that protected the town.[10] The fortress was a plain adobe structure with crenelations, without a moat or breastworks. In 1741 it was garrisoned by one "weak" company and armed with eight cannons. Most of Paita's inhabitants, except for a handful of local officials, some merchants, and the clergy, were Indians, *mestizos* (mixed bloods), or blacks.[11]

Paita's Hinterland

In the late colonial period the hinterland or *espacio económico* that Paita served was a rather densely populated agricultural region with several significant population centers other than Paita. A census of the *partido* of Piura - roughly equivalent to the modern Department of Piura - ordered by the first Archbishop of Lima, Mons. Geronimo Loayza, revealed a population of 193,000 in the mid sixteenth century. The demographic crisis of the first two decades of the conquest, which was much more severe in the coastal valleys than in the *sierra*, or highlands, had a devastating effect on the local population. By the late eighteenth century the population of the 80 or so towns, villages, hamlets, and haciendas of the *partido* of Piura, in the Intendency of Trujillo, had been reduced by epidemic disease to between 42,000 and 45,000. About 78 percent of the inhabitants of the *partido* of Piura in 1795 were Indians and *mestizos*. According to a contemporary description, they spoke a variety of languages "worthy of admiration." As a result of the population decline, the indigenous irrigation systems that Pizarro and his men admired for the first time in the Piura region had been abandoned, the canals collapsed and buried in sand. Nevertheless, significant demographic recovery had occurred by the mid-nineteenth century. The 1850 census gives the Department of Piura a population of 74,372, while the 1876 count shows 21,077 inhabitants for the province of Paita alone (of which 40 percent were classified as Indians) and 55,099 for the province of Piura, of which almost 72 percent were listed as Indians.[12]

The small irrigated farms of the valley of the Chira River, whose main channel was navigable by shallow-draft vessels year round for 20 miles inland, produced cotton, sugarcane, tobacco, indigo, and cochineal, in addition to table crops, fowl, and firewood for the markets of Paita and other towns of the Department of Piura. The Paita hinterland played a role similar to that of the *espacio económico* of Arequipa described by Brown in his *Bourbons and Brandy*.[13]

Economically, one of the most important of these towns was Colán, about 12 kilometers north of Paita, at the edge of a wide salt plain about 2 kilometers from the ocean, just south of the old mouth of the Chira River.[14] Colán, whose settlement also predates the arrival of the Spanish, had about 400 inhabitants during this period, almost all of them Indians. By the early 1830s, with the growing demand for water and provisions occasioned by the influx of Yankee whalers, the population had grown to about 2,000, according to one French visitor. Its church had been destroyed by an earthquake in the 1820s, but the old bell tower was a landmark for mariners navigating in the Bay of Paita. By the mid-1830s the parish church, dependent on the curacy of Paita, had been reconstructed and restaffed to minister to the spiritual needs of the growing population.[15]

San Miguel de Piura, some 40 leagues east-south east of the port, was the political and spiritual capital of the region.[16] On the eve of independence the town had a population of 7,000 to 9,000. A foreign traveler noted in 1812 that the town had a parish church, Franciscan and Mercedarian convents, and a hospital run by Bethlemite friars on the main square. The one-story dwellings in which Piura's inhabitants lived were constructed of adobe or *bajareque*, and the streets were unpaved. By 1833, according to another foreign visitor, the population of Piura had grown to 10,000.[17]

Southwest of Piura was another entirely Indian population center called Catacaos, which specialized in the manufacture of straw hats. Aside from hats, Catacaos's main claim to fame is having been the birthplace of one of Peru's most eminent educators and physicians, Cayetano Heredia.

The first seven leagues of the old road between Paita and Piura ascend gradually across the rocky, dusty *tablazo* to the post house or *tambo* at Congorá, where water and food brought in from the Chira valley were available at great cost. The second half of the fourteen-league trip, which took about six hours, descended through low, constantly-shifting, crescent-shaped sand dunes called *médanos*. Travelers generally left Paita at sundown, paused to rest for a few hours at Congorá, and continued on at two or three in the morning in order to arrive in Piura before the mid-morning sun made the desert unbearably hot.[18]

Paita as an Entrepôt

Paita's importance as a port of call during the colonial period was due directly to a force of nature, the Humboldt or Peru Current. This frigid stream of water, from 100 to

7

CHAPTER 1

150 miles wide, flowing north from the Antarctic along the coast of Chile and Peru as far as Parinas Point, where it veers out into the Pacific, moves at a velocity of about 32 kilometers (20 miles) every 24 hours.[19] Its chilling effect makes most of the Chilean littoral and all of Peru's coastal plain extremely arid and barren.

Moreover, sailing ships traveling from Acapulco, Panama, or Guayaquil to Callao, Lima's port, must buck the strong southwest winds that, during most of the year, accompany the Peru current on its progress toward the equator. These almost constant headwinds make it necessary for sailing vessels to repeatedly tack far out into the Pacific and then back toward land as they beat their way down the coast toward the City of Kings. This, of course, made the trip much longer and more difficult and costly, increasing the need to call at intermediate ports for water and provisions. Paita's ample and secure harbor, and the availability of all manner of provisions from nearby towns, made it a popular port of call for ships making the long haul south to Callao.

For large sailing ships making the voyage Guayaquil-Callao-Guayaquil, the round-trip often lasted from three to five months. An English traveler who sailed from Paita to Callao in 1812 made the trip in 51 days of difficult and tedious coasting. In contrast, a French corvette made the nonstop voyage from Callao to Paita in four days and nights in 1837. Ten years later a British vessel made it from Paita to Callao in over three weeks of "most tedious" sailing, but did the return trip in just five days.[20]

Regarding the difficulty of sailing south from Paita, the Spanish travelers Jorge Juan and Antonio de Ulloa related the following exaggerated anecdote in their eighteenth-century account:

The master of a merchant ship, who had been lately married at Payta, took his wife on board with him, in order to carry her to Callao. In the vessel she was delivered of a son, and before the

8

ship reached Callao, the boy could read distinctly. For after turning to windward two or three months, provisions growing short, the master put into some port, where several months were spent in procuring a fresh supply; and after another course of tacking, the same ill fortune pursued him; and thus four or five years were spent in tacking and victualing, to the ruin of the owner, before the ship reached Callao.[21]

The alternative to the long and tedious sea voyage from Paita south was to disembark at the port and continue to Lima overland. The trip by land from Paita to Lima, about 219 leagues, could be made in 18 to 20 days along a "tolerably good" coastal road. Although most of the road traversed desert terrain, including the formidable Sechura plain just south of Paita, there were oases, villages, and adequate post houses all along the way. Describing a trip from Lima to Trujillo and on to Piura and Paita in 1812, Stevenson noted how the "most luxurious vegetation" of the river valley oasis "clothed in the...garb of spring or autumn" enlivened the spirit of the weary traveler.[22]

Given the option, most travelers bound for Callao chose to disembark at Paita and continue on to Lima along this coastal road.[23] Some preferred to make the trip on mule back, carrying with them all the supplies needed for the journey. For the illumination of English travelers who might find it necessary to venture "any considerable distance" from Lima, Stevenson described the bedding, equipment, and supplies that would be necessary to take along. He also revealed his strategy for procuring free lodging in the towns and villages along the way, noting that in most places the local curate provided the best table and "society."[24]

Other travelers making the overland trip from Paita to Lima, particularly women, preferred to travel in litters. These were square wooden boxes with curtained openings on each side and a pad or mattress fitted in the bottom. The litter was suspended from two long bamboo poles, one on each side, which were strapped to the saddles of two mules. A young boy often rode astride the lead mule to act as a guide. Stevenson disliked traveling by *litera*, since the elasticity of the bamboo poles combined with the uneven stride of the two mules caused it to bounce up and down and roll from side to side, producing "all the effects of seasickness, besides a universal soreness in the limbs."[25]

Most of the thousands of mules that were used to carry passengers and freight on the Paita-Lima route came from the area around Piura. These animals were also used extensively in the circular trade between northern Peru and the area of southern Ecuador around Loja, in which the Peruvians exchanged raw cotton for cloth known as *tocuyos* and *bayetas*. According to Stevenson, most of the men of Piura were mule breeders or muleteers, and a prime Piura mule could be worth the "amazing" sum of 250 pesos.[26]

It is clear, then, that even before the arrival of the New England whalers at Paita or the subsequent introduction of steam-powered ships on the Panama-Callao run, Paita played an extremely important role as a port of call and source of provisions. It was even more important as an entrepôt for the storage and transfer of south- and north-bound goods, and as a point of disembarcation for Lima-bound passengers who wished to avoid the delay and tedium of a long sea voyage. If customhouse and immigration records for Paita during the colonial period existed, they would provide eloquent testimony to the port's role as an outpost of empire on the northern coast of Peru.

Paita as Prize

Paita was never a physically attractive place. But it was a useful place, an important place within the Peruvian context. One measure of its importance was the frequency with which it was attacked and pillaged by foreign pirates. Sir Thomas Cavendish sacked the port in 1557, and Dutch corsairs attacked and burned Paita twice in the seventeenth century. The crew of the 16-gun *Cygnet* burned the town in

1604 after their demands for a ransom of flour, sugar, wine, and water were rejected, while in 1720 Captain George Shevlocke of the *Speedwell* and 60 men burned Paita after the Spanish authorities refused to ransom the town for 1,000 pesos.[27]

Lord George Anson (1697-1762), whose squadron of five ships attacked Paita in 1741, published a detailed account of his adventures in 1748.[28] The English stole from the customhouse a shipment of silver bullion that was about to be shipped to Acapulco to purchase goods from the Manila galleon, and ransacked Paita's "many valuable warehouses full of valuable effects." Private homes were pillaged, 30,000 pesos in church plate, sterling, and jewelry was stolen, and hogs and fowl in great numbers were commandeered for the crew's mess. After spiking the cannon in the fort, the English torched the town's highly flammable houses and the squadron sailed away (see plate 3). Spanish officials later claimed that 1.5 million pesos worth of goods were stolen or lost in the fire that engulfed Paita. The loss would have been even greater if the governor of the port, who had been forewarned of the English squadron's approach, had not moved the royal treasury and his own fortune inland to Piura before Anson's arrival.[29]

After leaving Paita, Anson's flagship the *Centurion* met up with a sistership, the *Gloucester*, which had taken two prizes enroute from Callao. One ship yielded large quantities of wine, brandies, and olives in jars, plus the equivalent of £7,000 in cash. The second, a large launch, was laden with cotton packed in clay jars, which were taken aboard the *Gloucester*. The surprised passengers, who were dining on "pidgeon [sic] pie served up in silver dishes" when the launch was seized, also were taken aboard. The clay jars were later found to contain £12,000 in "double dubloons" and pesos destined for the merchants of Paita.[30]

In April of 1819 another English squadron commanded by Lord Thomas Cochrane attacked and looted Paita. In his *Memoirs* the British legionaire William Miller, who participated in Cochrane's raid, called Paita "little Jamaica," a "grand entrepôt for contraband" from the West Indies across the Isthmus of Panama.[31] While none of these attacks by foreign adventurers can be considered a great strategic or military victory, given Paita's miserable defenses and its isolation, the detailed description of the booty taken during the two British attacks on Paita clearly shows how important it was as a port of entry for the viceroyalty and as an entrepôt for the trade between Panama and Callao.[32]

Paita on the Eve of Independence

Despite the paucity of historical records, our informed imagination could paint the following portrait of Paita in the late colonial period. Four or five 300-ton Spanish merchant vessels and perhaps a royal ship of the line are anchored in Paita's tranquil and ample bay. One of them is undergoing repairs and refitting. Balsa wood rafts, dugout canoes, and small skiffs ply the bay between shore and ship, carrying water, provisions, passengers bound for Panama, and - once or twice a year - perhaps a shipment of silver bullion from the Lima mint. The same small craft are employed to carry ashore boxes, barrels, and trunks full of silk, porcelain, and spices brought by the Manila galleon to Acapulco. They might also carry European goods that had been transported across the Isthmus of Panama from the fair at Portobelo, and perhaps a newly-appointed royal official, a high clergyman, or a Spanish merchant about to begin the overland trip to the capital of the viceroyalty.

A typical cargo of a Spanish ship just arrived from Panama and anchored in the bay at Paita might resemble that of the *Nuestra Señora del Carmen*, a 270-ton vessel with a crew of 43 that was captured by Anson off Paita in 1741. The cargo seized by Anson consisted of iron, steel, wax, black pepper, cedar planks, snuff, rosaries and other "Romish indulgences," cinnamon, indigo, and European textiles. Another typical cargo for a ship bound for Callao from Guayaquil during the same period was that of the 300-ton *Santa Teresa de Jesús*, which contained timber, cacao, coconuts, tobacco, hides, Quito cloth, and wax.[33]

By day the narrow, dusty, unpaved streets of Paita teemed with ships' officers transacting official business or negotiating the purchase of provisions. Indian and *mestizo* muleteers from Piura herded their animals laden with baled cotton, salt, indigo, and soap toward the market square or the houses of local merchants. Women selling watermelons and other fruit sat in the welcome shade of a house or under a cotton parasol crying their produce. Fishermen just returned from the bay hawked their catch, and everywhere dusty, barefoot children frolicked amid the throngs.

The governor's house, the customhouse, and the fort were bustling with official activity, while in the parish church or the church of La Merced a mass was being sung to thank God for a safe sea voyage or to implore His blessing on a trip about to begin. If the fleet brought in from Panama or Acapulco luxury goods for the sophisticated tastes of the Lima aristocracy, the agents of one of the capital's powerful merchant houses might be lodged in one of Paita's modest inns or in the house of a relative or business associate, from whence he would negotiate with a ship captain for the outright purchase of broadcloths, silks, cambrics, and velvets. Or perhaps he would agree to trade them for Peruvian wine, rum, and brandy stored in one of Paita's numerous warehouses.[34] If a load of bullion had arrived recently from Lima for transshipment to Panama and Madrid, the customhouse would be heavily guarded by troops of the local garrison.

At night, under that brilliant moon for which Paita is renowned, the port's dimly lit *pulperias, fondas,* and *cantinas* echoed with the click of billiard balls and the clink of glasses.[35] Somewhere someone played a guitar and sang a plaintive love song or a festive *jarana*. Many of the town's residents - almost all *mestizo*, Indian, and black - lolled in

CHAPTER 1

hammocks or sat in their doorways, catching the cool evening breezes that blow in across the desert, smoking and gossiping about the latest news from Lima or the arrival of some dignitary from Panama. Sailors from a merchantman or ship of the line, and soldiers from the garrison - dressed in their finest - strolled from cantina to cantina, their eyes and ears primed for the subtle suggestion of excitement or romance (see plate 4).

2

The Foreign Presence in Paita

G eneral Miller's reference to Paita as "little Jamaica" clearly underscores the fact that, even before Peru's formal independence from Spain in 1821, royal control over the import of non-Spanish goods into Paita and over the arrival of non-Spaniards into that remote corner of the empire had lapsed. In part this was due to a dramatic change in imperial policy, the decision in 1778 to liberalize trade between the Spanish American colonies and the nations of the North Atlantic. But even before trade liberalization was officially sanctioned, because of its distance from the center of viceregal authority in Lima, Paita probably served as an important conduit for the importation of illegal European goods long before the final curtain was drawn on Spain's domination over Peru. The very nature of the contraband trade, however, makes it difficult or impossible to document this assertion.

It is possible, however, to document the presence of non-Spaniards in the northern reaches of the viceroyalty as far back as the middle of the eighteenth century. As we have seen, the Englishman Charles Stevenson visited Paita as part of his extensive travels in the viceroyalty. Walters reported that a "papist" shipmaster named Gordon was the captain of the civilian mounted militia that ineffectively tried to defend Paita from Anson's assault in 1741. An English ship carpenter from Portsmouth who was employed in the Guayaquil shipyards, and for whom Anson was carrying letters from home, died defending Paita against his countrymen's assault. Anson also encountered an itinerant Irish peddler named John Williams in Paita. By his own account, Williams had lived for years in Mexico and made a lot of money, but when Anson found him in 1741 he was penniless, in rags, and had just been released from the Paita jail.[1]

Similarly, in his travels in Peru in 1823 and 1824, the Englishman Robert Proctor met a countryman who had lived in the Spanish realms of South America for 30 years. A baker by profession, he first came to Peru via Panama as a clown in an English carnival company, and lived in Quito, Cusco, Arequipa, and La Paz (where he married). Proctor met him in Pativilca in 1824, where he kept a "hucksters" shop and was known locally as Julian Campos, because his mother's maiden name was Fields.[2]

Of course Paita's non-Spanish residents and visitors never approached the numbers of the foreign populations of Rio de Janiero (3,000 plus in 1820) or Buenos Aires (3,700 plus in 1822) during the wars of liberation.[3] But once formal independence was achieved, the port's economic vitality soon attracted significant numbers of foreigners. By 1830 the majority of these foreigners were North Americans, most of whom were the officers and crews of the intrepid Yankee whaling ships that ranged the Pacific in search of sperm whales.

THE FOREIGN PRESENCE IN PAITA
The Arrival of New England Whalers

The history of the New England sperm whaling industry from the 1790s until the American Civil War has yet to be told in a comprehensive, systematic way, even though the primary sources for such a tale are abundant.[4] A summary review of the development of sperm whale fishing in the South Pacific will serve to introduce the discussion of the economic and social impact of whaling on Paita.

Although American whalers were active in right whale fishing in the Atlantic early on, the first recorded catch of a free-ranging sperm whale is credited to a Nantucket vessel in 1712. The waxy oil and spermaceti of the sperm whale, which preferred the warmer waters of the South Atlantic and the Pacific, were worth far more than right whale oil.[5] In the 1760s, tryworks, brick rendering furnaces for the reduction of whale blubber, began appearing on American vessels.[6]

The first Massachusetts vessel to double Cape Horn was the 240-ton ship *Beaver*, commanded by Captain Paul Worth. She left Nantucket in August 1791, and early the following year rounded Cape Horn to enter the Pacific. The first American vessel to return from a successful Pacific whaling cruise was the *Rebecca* of New Bedford, which departed shortly after the *Beaver* but returned in February of 1793. Calculated in the standard 31.5-gallon barrel whale oil measure, the *Beaver*'s cargo totaled 1,100 barrels of sperm oil, 370 barrels of head matter, and 250 barrels of whale oil, making a profit of £20,000 on this, her maiden voyage. In other ways, however, Captain Worth's experience off the Peruvian coast was not an auspicious beginning for the Yankee whalers who would, in subsequent years, frequent those waters. The *Beaver* was boarded by a Spanish ship of war off that coast and warned not to put into any port in the viceroyalty of Peru. In need of fresh provisions, however, Captain Worth entered Callao, only to be driven out with a strong warning that he and his vessel would be seized if he returned.[7]

Inspired by the success of the *Beaver* and *Rebecca*, Captain George Bunker took the whaleship *Washington* into the Pacific, reaching Callao in July 1792. The Spanish authorities did not allow Bunker to land, but they permitted him to take on board fresh water, potatoes and other vegetables, and fresh meat. The first recorded North Atlantic whaler to call at Paita, the *Emilia* of London, arrived on the coast about the same time. Unfortunately, the journal of the *Emilia*'s young Scottish cooper, John Nicol, contains no description of the northern port.

According to a comprehensive report on New England whaling in the Pacific, prepared by the American consul in Paita and dated 1858, sustained American whaling off the west coast of South America began after 1812, when Captain David Porter took the U.S. frigate *Essex* into the Pacific to destroy British whaleships. During the War of 1812 there were only 10 or 12 American whalers in the entire Pacific, and after the conflict the New England whaleman "found this immense space entirely at his disposal."[8]

In July of 1815 *Niles Weekly Register* reported that some 16 whaling vessels had sailed from eastern seaports for the Pacific, and that the pace of the industry was picking up.[9] Four years later, despite a general decline in United States commerce, *Niles* reported that a great deal of new capital was being invested in Pacific Ocean whaling, "a great nursery of seamen." By August of 1819 some 60 New England whalers were working the Pacific fishing grounds.[10]

The impressive growth and technological advancement of the New England-based sperm whale fishing industry continued to merit comment by the North American press. By early 1820, 63 United States-owned or -commanded vessels were engaged in sperm whale fishing in the Pacific.[11] Late in the following year 41 whalers from Nantucket alone were engaged in fishing off the coast of Peru and Chile and off the Galapagos islands.[12]

In mid-1823, *Niles*, citing the *Nantucket Inquirer*, noted that for the period 1821-22, "comprising the average term of one sailing voyage," approximately 100 vessels based in Nantucket and New Bedford had off-loaded some 4,360,000 gallons (about 138,400 barrels) of whale and sperm oil. In late 1823, again quoting the Nantucket paper, *Niles* noted that the number of Nantucket whalers in the South Seas was more than double that of English and French whalers, and that most of those were commanded by American masters. The quantity of sperm oil alone imported into Nantucket and New Bedford for the years 1820, 1821, and 1822 was 998,804 barrels, 1,336,987 barrels, and 1,172,398 barrels respectively.[13]

By 1830 New Bedford was home port for 120 square-rigged vessels (see plate 5), mostly whalers. In that year 41,144 barrels of sperm oil were imported into New Bedford alone, slightly less than the volume of Atlantic-origin right whale oil shipped that same year. The following year, the New England whale fleet consisted of about 290 ships and barks - 170 sperm whalers and 120 right whalers - whose annual consumption of copper amounted to 700,000 pounds.[14] As an approximate point of comparison, in 1832 - the first year for which consular returns for United States ship calls at Paita are available - those 64 ships that put into the Peruvian port carried 40,895 barrels of sperm oil.[15]

In response to the Pacific whaling boom, the population of New Bedford burgeoned from less than 4,000 in 1820 to 7,592 in 1830, 11,113 in 1836, and over 20,000 in 1855. The city boasted ten spermaceti candle factories in 1832. Six years later there were 17 candle factories and oil manufactories in New Bedford. By the beginning of the United States Civil War, over 12,000 men and women were employed - on and off shore - in the New Bedford whaling industry, which generated annual revenues in excess of

CHAPTER 2

$12,000,000. There are no satisfactory treatments of the macro-economic impact of the whaling industry on the growth and development of New England and of the United States in the nineteenth century. It seems clear, however, that the capital generated by whaling financed the construction of railroads in Massachusetts, the textile industry in New Bedford and Fall River, ironworks in Pennsylvania, the lumber industry of Michigan, and the nascent petroleum refining industry in Pennsylvania.[16]

The golden age of the New England whaling industry in the Pacific began in the mid-1830s and lasted for approximately ten years. In June 1835 the acting United States consul in Paita reported that the number of American whaleships calling at Paita exceeded that of any previous year. According to the *Sailors Magazine and Naval Journal*, the same year 88 whaleships called at Paita, along with 21 American merchant ships. Ten years later, 157,917 barrels of sperm oil and 272,730 barrels of right whale oil were produced by the New England whaling industry, representing increases of almost 50 percent and slightly more than 300 percent, respectively, over the volumes for 1830.[17] By 1846 some 735 ships, barks, brigs, and schooners - most of them from Nantucket and New Bedford - plied the world's oceans in pursuit of whales. Their aggregate tonnage was some 233,200 tons, and their cargo in that year was valued at $21,075,000.[18] In 1846 the United States consul in Paita reported that over 400 American whalers were cruising the Pacific.[19] By the following year, however, the decade of impressive growth had so over-stimulated the industry that production began to exceed demand, and prices fell somewhat.[20]

But the importance of this phenomenal growth of New England whaling did not lie solely in numbers. The pioneer historian of the Yankee whaling industry, Alexander Starbuck, perceived its importance thusly. To Starbuck the whalers were "the pioneers of the sea,...the advance guard, the forlorn hope of civilization." In his eth-

20

nocentric view, they were the vanguard of North American influence in the Pacific. "The ports of the western coast of South America," he wrote, "first beheld the Stars and Stripes as the standard of [an American whaleman]. In their footsteps Christianity [sic] followed, into the field opened by them flowed the trade of the civilized [sic] world."[21]

Surviving whaleship logs and journals reveal a pattern of visitation at Paita. Typically, an outwardbound New England whaler would stop at the Azores or Cape Verde Islands in the Atlantic for supplies before rounding Cape Horn and seeking sperm whales along the Chilean and Peruvian coasts. Then they would call at Tumbes for wood and water and Paita to buy potatoes and onions, to collect mail, and to give their crews liberty on shore. From Peru, the ships headed west, whaling along the equator and often heading into the Sea of Japan and up into the North Pacific or down to the grounds around New Zealand before calling at the Sandwich (Hawaiian) Islands to resupply. Some ships then headed east to whale along the Mexican and Peruvian coasts, again calling at Paita, and some made a final call there before heading round Cape Horn on their return to New England. As an example, the New London, Connecticut, whaleship *Chelsea* stopped at Paita on 14 January 1829, 18 months into a 44-month voyage. "Came to anchor in Payta Lay there 17 hours got potatoes onions & pumpkins &c then weighed anchor & went to sea," recorded the ship's logbook.[22]

The early American whalemen might have reacted to Paita as did Navy Surgeon Thomas Boyd of the U.S.S. *Brandywine* in July of 1827.

This is without exception the most uninviting desolate spot that ever human beings selected for a habitation, and every thing which denotes their appearance instead of enlivening but increases the rude & I might say unnatural features of nature. - The Bay seems to have been formed by a sudden disappearance of a portion of a high table land, which has left a perpendicular bank of white clay around the harbour. The water has here & there left a strip of

CHAPTER 2

sandy beach: one of these spaces about two hundred cane constructed huts are located - This is Payta - A few miles farther on the margin of the sea under a sand bank is another village called Colan -

From the ship there is not a living thing or circumstances which might induce us to believe there were living things to be seen not a tree or shrub - nothing but a barren sand waste - apparently incompatible with life or verdure - The huts of the village look more like pilgrims temporary shelter from a burning sun than places of habitation -

After a visit on shore I have but little to say in favour of Payta as a community or place - The huts are composed of cane wattle together supported by post & covered with the leaves of the cactus, without floors & but seldom divided into apartments - The furniture was as simple as the buildings consisting of a bench, a few large gourds & a hammack stretched across - generally in a corner there was a small fire over which hung a pot, the sole cooking utensil their house-wifery skill requires - The inhabitants of these rude dwellings...are nearly unmixed Indian. Here & there a white skin or curly hair indicate an acquaintance with the European and African race but the mixture is comparatively rare.

There are few of the features, as may be supposed, of a regular organized society in Payta - Their simplicity, indolence & benevolence are not excelled by any people. The former is almost patriachal - We went to see the Padre who is always the most considerable man in small places: he was surrounded with nephews & grand nephews, neices & grand neices who would have been perplexed to have found a father if required - Some were in the cool refreshing state of nudity so admirably adapted to the climate that we scarcely noticed it as an impingement on delicacy, others were clothed but in the extremity of a fashion which once obtained in our country to aid consumptions and concupisence - We have often occasion to remark how extremes touch upon each other - The licentiousness & fashions of fashionable refined society and the unrestrained indulgences & customs of nature are so nearly alike as to be merely the reflexions of each other. The good Padre welcomed us to his house with as much cordiality as if it had been a palace, admired every thing about our persons, & wished to buy while his younger family begged vehemently -

THE FOREIGN PRESENCE IN PAITA

We stroll'd through the village, entered the huts without ceremony and were uniformly kindly received. The original Indians acted differently from the Padre; they manifested neither curiosity nor desire and seemed to consider themselves in possession of all that was necessary for happiness, as they regarded our situation with pity whenever we wished to know any thing of their customs & habits considering it as a mark of their superiority.

Fighting game fowls is the only amusement which calls forth their excitement. They are passionately fond of it. In every hut one or two which had no equivalent in the form of money -

There is not a drop of fresh water within twelve miles of Payta. It is brought on jacks in large gourds or jars which are carefully kept under lock....

We are in the winter season of this climate and for purity, equability & healthfulness nothing can be conceived more delightful - The sun is warm at noonday but it raises no vaporious exhalations, a constant breeze brings the refreshment of the sea air - The thermometer generally ranges between 70 & 76 at this season - It never rains.[23]

Although seasonal in nature, the impact of the whaling boom of the 1830s on Paita was immediate and visible (see plate 6). A sailor aboard the ship *Chelsea* of New London, who visited the port in 1831, estimated the population at about 3,000, a figure corroborated by an American navy doctor whose ship called at the port two years later. Six priests, one of whom had an angelic face and the reputation of "a complete gallant," tended to the town's spiritual needs. Whereas the 1831 visitor found no house in Paita fit for a stable, the navy doctor noted that new houses were being built along the main street, while older buildings were being refurbished as a direct result of the economic activity associated with the arrival of growing numbers of American whaling ships. The lithographs from de la Salle's 1837 account, which depict many two-story houses, some with plastered facades, balconies, and barred windows, attest to the prosperity brought to Paita by the New England whalers (see plate 7). While many of the houses were still like "chicken cages," a French traveler noted, some were quite luxurious on the inside (see plate 8).[24]

When the New Bedford whaleship *Mercury* called there in October of 1837, many of Paita's residents had fled to the interior, fearing attack by a Chilean fleet. Still, the ship was able to obtain onions, potatoes, and beef, and a whaleman noted: "the port contains about one thousand inhabitants - the houses are built of Bamboo fill'd in with mud or plaster - no floors but the ground - and the roofs are thached with flags or rushes - scarcely any furniture or cooking-utensils to be seen in any of the houses - it appears to be a complete abode of wretchedness and poverty - although there are some wealthy people here."[25]

Thus, even before the "boom" decade of 1835-46, Paita had become an important port of call for American whalers to take on supplies; to refresh their crews; to receive mail from home; to release sick or injured crew members; and to cooper their oil. For those crewmen fortunate enough to get shore leave - and whaleships usually remained long enough to allow each watch (half the crew) or each boat's crew (six men) at least one day of liberty ashore - a "fashionable resort" called the *Unión Sociable* offered billiards and coffee. The American doctor described the crewmen from the whalers "swaggering before the doors of the *pulperías* and talking of their exploits with 'the fish.'" On summer evenings the town's entire population seemed to live in the streets, playing guitars, dancing, and engaging in "innocent amusements" - and others perhaps not so innocent. An American woman who stopped in Paita for several days in 1868 wrote that "the moonlit evenings are indeed enrapturing."[26]

A French visitor in 1836 estimated the population at 4,000, and noted the presence of several prosperous merchant houses whose annual trade totaled some eight to ten million pesos.[27] A countryman observed four years earlier that "money is the god of the Europeans who have been

24

transplanted to this miserable village," whose ignorance and stinginess he described as repugnant. The contraband trade continued to flourish. By the mid-1840s Paita boasted about 800 houses, but a British traveler found the public buildings "small, and undeserving of public attention."[28] Small wonder, in view of the wanton depredations of most of Paita's foreign visitors during the colonial period.

Paita as a Service Economy

Although rest and recreation for sea-weary crewmen was important, as numerous contemporary accounts attest, Paita played a far more important role as a source of water, food, provisions, firewood, and marine stores for the North American vessels that called there. The fact that Paita was totally destitute of its own source of fresh water did not prevent the port from playing a key role in supplying this essential element to the American fleet. Beginning in colonial times, fresh water from the Chira River was ferried across the bay on a daily basis from the Indian community of Colán on large balsa rafts. According to an early eighteenth-century account, the Indians of Colán took advantage of the diurnal land winds to coast down the bay to Paita on their rafts, returning at night with the sea winds behind them. The water was carried in large earthenware jars and sold "at very high price" to residents of Paita and to shipmasters.[29]

By the 1820s mules had replaced the balsa rafts in the transport of drinking water to Paita, but it was still expensive - "as much so as the wines of France," according to one French visitor (plate 9). The American consul reported in 1854 that in normal times water from Colán sold for two cents a gallon, but that when Peruvian warships called at the port, speculation drove the price up. He commented that an American merchant house in Paita had proposed to the Peruvian government the building of a dam and an aqueduct to channel water from the Chira River into Paita.[30]

CHAPTER 2

Apparently nothing came of the proposal, however, since in 1861, 1862, and 1877 three different foreign observers noted that water was still being brought in on mule back from the Chira and that it was still expensive, at three to four *reales* for a 24-gallon jar. One sanctimonious Yankee doctor observed that although the port's "natives" might get enough water for drinking, "it is *palpably* [his emphasis] too costly for cleanliness." Sometime after 1865, probably as a result of flooding associated with the El Niño phenomenon, the Chira created a new mouth about five kilometers north of the delta at Colán, distancing the town from the river and reducing the comparative advantage of the residents of Colán in the water trade with Paita.[31]

In contrast to the dearness of drinking water in Paita, food was fairly cheap. A wide variety of native and European food crops were cultivated on the small farms owned by Indians and *mestizos* in the lower Chira valley. They included onions, potatoes, carrots, cucumbers, caulifower, yams, sweet potatoes, yucca, achiras, corn, dates, coconuts, melons, pomegranates, citrus fruit, guavas, guayabas, mangos, tunas, and tamarinds. Wheat came from Lambayeque or from Chile, and beef, pork, poultry, fish, and kid were available to ships' cooks at reasonable prices. An American sailor who visited Paita in 1831 noted in his journal that the night before disembarking he spent the whole night thinking of the fresh fruit, bread and butter, and fresh beef that awaited him in the morning. Some provisions were delivered right to the ships by balsa rafts, while others were available in the market. Even as late as 1905 one American traveler commented on the incredible number of goats to be seen in and around Paita.[32]

In addition to water and food, the American whalers also took on other important provisions at Paita. Large quantities of firewood for cooking and to fuel the tryworks were brought in from the Chira valley and from Piura on mule back and sold in Paita (in 1861) for two *reales* a load, equal to about a dozen arm-sized sticks of *algarrobo* or carob wood (plate 10). Although it was considered expensive, the algarrobo's hardness and high caloric content made it excellent firewood.[33] Salt, which was essential for cooking and for curing meats, was produced by Indians who lived at the mouth of the Piura River on the shore of Sechura Bay, south of Paita. It was transported by balsa rafts to Paita for sale to the fleet.[34]

Another essential commodity acquired by the whaling fleet at Paita was soap. The trade in soap made from goat tallow and in tanned goat hides (*cordovanes*) was one of the north coast's principal economic activities throughout most of the colonial period.[35] Stevenson noted soap works in Piura in 1812. They employed goat tallow to produce large quantities of soap and exported cordovan goat hides to cities from Lima to Panama as a by-product. He also observed a soap manufactory in operation in Lambayeque, to the south, where large herds of goats grazed on carob beans. Alkali from the ashes of the *lito* shrub, found in abundance in the Sechura and Lambayeque valleys, was boiled with goat tallow to produce a very hard soap, which a British traveler judged to be inferior to English soap.[36] Ruschenberger also observed a rather primitive soap works at Chiclayo in 1833. The soap was transported from Piura to Paita by mule or from the south by balsa raft, where it was sold to the American whalers and other vessels.[37] Aldana provides a fairly complete series on the price of soap in Piura beginning in 1590. For the period 1810-31 the average price was 22 pesos 3 reales the *quintal* (hundredweight), whereas in Lima, in 1825, the price was around 40 pesos for the same measure.[38]

A whaler that had been at sea for five or six months needed other goods and services besides the water, food, firewood, salt, and soap that Paita provided. Equally important, especially for the older ships employed in the Pacific whale fisheries, were marine stores and minor repairs. Many of the ships that called at Paita were venerable and needed constant maintenance. They included the *Golconda* built in 1807, which called at Paita 13 times between 1835 and 1862.[39] Others were the *Hesper* of Fairhaven (1811), condemned and sold in Paita in October 1863 after having called there a dozen times; the *China* of New Bedford (1816); the aptly named *Balaena* and the *Hector* (1818); the *Roscoe* of Fairhaven (1821), and the *Hydaspe* of New Bedford (1822).[40] All kinds of services and naval supplies, many of which were of local manufacture, were readily available from Higginson & Company, Ruden & Company, or one of the other merchant houses doing business in Paita.[41]

The square-riggers of the New England whaling fleet continually needed to patch and replace their sails, even in the relatively benign sailing conditions that prevailed in the Pacific. There were three basic sources for the sailcloth that the whalers bought in Paita. Some second-hand sails, generally of American manufacture, were available from other whaling ships, from American merchantmen, or from ships that had been dismantled or scuttled in Paita. New American sailcloth was also available from merchants with headquarters in Callao, including the firm of John Bryce & Company (successors to Pablo Romero) where the Irish-American entrepreneur William R. Grace got his start in the early 1850s.[42]

For the purposes of this study, however, the consumption of locally manufactured cotton sailcloth by American whalers and the stimulus it gave to the local economy is a far more interesting phenomenon. Clayton, in his study of the Guayaquil shipyards, mentions that cotton sailcloth loomed in Chachapoyas and Cajamarca, in northern Peru, was thought to be comparable in quality to European goods.[43] In 1807 the looms of Piura produced 30,000 *varas* of canvas (a *vara* is 33 inches, or 84 centimeters). When Stevenson visited Lambayeque in 1812, he listed hand-spun and handwoven cotton canvas among the town's principal manufactures.[44] This coarse cotton cloth was probably produced in the many Indian *obrajillos,* and home workshops that sprung up in Peru in the late colonial and early republican period, whose simple family-oriented labor mobilization and low overhead helped them to survive the onslaught of cheap imported cloth in the 1820s.[45]

The American consul noted in a report to the State Department in 1854 that coarse cotton sailcloth, generally used by the smaller vessels, was still being loomed in the department of Piura.[46] None of the available evidence even suggests the relative importance of these several sources of sailcloth in any given year or over time. It is reasonable to assume, however, that as the pace of whaling, trade, and coastal navigation increased along the Peruvian coast, the better quality imported canvas prevailed over the local product. Nevertheless, during a twelve-month period around the year 1890, over 4,000 kilograms of cotton canvas worth $2,495 was produced in the Piura region and exported through Paita, probably to the coastal trade.[47]

The American whalers and other sailing ships that called at Paita also needed to replace the lines and rigging that characterized this kind of vessel. Here too, over time, imported rope and cordage probably replaced local products, but there were a number of indigenous sources that could be drawn upon in the early decades of the nineteenth century. In the area around Guayaquil, whose shipyards generated a considerable demand for all kinds of naval stores during the colonial era, cordage was made from the fiber of the cabuya plant. The fibers were soaked, dried, and twisted into rope and then treated with pitch for use on larger vessels. Clayton mentions a cordwainery working on Puna Island in the Gulf of Guayaquil as early as 1562.[48]

Two other Latin American sources for cordage were the cities of Léon and Granada in the Captaincy General of Guatemala (Nicaragua), where henequen or *jarcia* was twisted into a serviceable rope, and southern Chile. Chilean cordage was made of hemp or *cáñamo*, which was deemed superior to hemp from northern Europe because of its longer fibers. During the colonial period and into the nineteenth century rope manufactured in southern Chile was used extensively along the entire Pacific coast from Concepción to Panama.[49] Rope was even manufactured in the Intendency of Piura in the early nineteenth century. Stevenson mentions that, in 1812, many people in the interior of the province were employed in making rope from maguey fibers. The maguey, or *pita*, grew abundantly in the piedmont region of the *partido* of Piura in the late colonial period. Cordage made from *pita* was used to tie up bales and secure cargoes on balsa rafts and canoes, but according to the British observer was not used in fitting out larger vessels, probably because of its inferior tensile strength.[50]

Tar and pitch for caulking and for waterproofing rigging were key stores that were also available locally. Natural tar springs from the Amotape range, or Cerro de la Brea - which takes its name from the village of Amotape on the north bank of the Chira River, and extends northwest to the border with Ecuador - were the nearest source for Paita. The tar, known locally as *copé* or *copei*, was almost as hard as asphalt and was mixed with pitch to caulk vessels. Tar and pitch were also found near Guayaquil, in the *pozos de copei* on the Santa Elena peninsula, and in Mexico. For caulking the hulls of wooden ships, dried shredded coconut husks from Ecuador were mixed with pitch to make oakum for use below the waterline, because the fiber expanded when wet and resisted rotting. Hemp and other fibers were used to caulk above the waterline.[51] In addition, tar from Talara was used extensively to seal the interiors of earthenware jars and animal-skin bladders used to transport liquids.[52]

There were no local sources for the iron needed for nails, chains, anchors, and hardware; or for the copper used for sheathing the hulls of vessels to protect them from ship worms and for making ladles and other utensils used in the extraction and packaging of sperm oil. During the colonial period and well into the nineteenth century iron bars were imported from Europe across the Isthmus of Panama. Latin American iron, mostly from Mexico, was too brittle for naval applications. With the advent of the whaling boom in the Pacific, New England vessels brought increasing quantities of American iron for the maritime needs of Paita. Much of it was recycled.[53] Copper was obtained from Chile, but was frequently processed in Lima.[54]

Finally, the wooden planking and spars needed to refit and repair the whalers that called at Paita came generally from Guayaquil or were brought directly from the United States on board the whaleships themselves. In the mid-1850s there were no shipwrights resident in Paita. Nor were there lumberyards for ship timbers in Paita, or shipyards or drydocks for the repair of large ships. When whalers put into port for repairs, the work was usually done by crew members or local carpenters. According to several sources, the cost of labor in Paita was the lowest of any port in Peru.[55] Thanks to the absolute calmness of the bay, it was frequently possible to undertake extensive repairs while the ship remained afloat.[56]

Another factor that must be considered in this examination of the goods and services provided by Paita to the New England whaling industry is the number of Peruvian seamen who shipped on board Yankee whalers during this period. There is no single, comprehensive source to indicate how many Peruvian sailors shipped out of Paita on United States ships. During the early period, the crewmen were all Americans. Then, according to Starbuck, Afro-Americans, Portuguese from the Azores, "Kanakas" from the Pacific Islands, and "mongrels" from the western coast of South America began to replace the New Englanders.[57]

Although, according to one contemporary source, Peruvian seamen preferred to ship aboard English or Mexican vessels, the numbers signing on American vessels increased toward the middle of the century, and many of them came from Paita. In proportion to its population, Paita furnished more seamen for foreign ships than any other Peruvian port, except perhaps Callao. The evidence suggesting the percentage of Spanish American crewmen on New England whalers during the declining years of the Pacific whale industry is very fragmentary. During the first quarter of 1869, for example, of the 16 mariners shipped or discharged in Paita under the consul's authority, nine had Hispanic surnames. Similarly, during the last quarter of the same year, of the 37 sailors listed in a similar document, 13 were Hispanics, and most of them signed aboard American ships in Paita. Another indication of the growing importance of Peruvian hands on board American whalers, beginning in the early 1850s, is the regular purchase of large quantities of sweet potatoes (*camotes*) in Tumbes for the ships' messes. *Camotes* were and still are one of the staples of the working-class Peruvian diet.

Peruvian sailors were paid $15 a month, plus a sign-up bonus of $16. Since they signed on for a specified period of time rather than the duration of the voyage, the Peruvian sailors did not receive a "lay" or share in the oil cargo. In contrast, the American whalemen received no wages but signed aboard for a predetermined "lay," say 1/150th, which represented their portion of whatever the cargo was worth upon the ship's return. The Peruvians' wages of fifty cents per day would equate to a fair lay of a good voyage or a good lay of a fair voyage. Also, as foreign nationals employed on an American ship, the Peruvians had little recourse in case of abuse. They did not enjoy the protection of the American consul.[58]

It would be gratifying to be able to identify the Paita merchants who provided provisions, naval stores, and services to the New England whalers, to shed light on the origins of these provisions, and to establish approximate annual values for the goods and services provided to the fleet. But in the course of my research I have found only scattered, representative documentation that would allow me to even hazard a guess in this regard (see chapter 5). One can only infer, on the basis of the numbers of United States ship calls recorded by the consulate at Paita during the period 1832-65, the general dimensions of this trade and its economic impact on Paita and its hinterland.

Prostitution in Paita

This discussion of the goods and services provided by Paita and its hinterland to the American whaling fleet, along with their economic impact, would be incomplete without a reference to the oldest of professions. Yet the silence of the consular records that I consulted on this very important facet of Paita's role as a bridge between two economies and societies is lapidary.

Logic and common sense tell us that a port where - on the average - 45 American whalers and merchant ships called each year for a third of a century, each ship with an average crew of 25 or 30, must have had a substantial and active community of prostitutes.[59] Even if only a third of the total number of crew members were given shore leave during port calls at Paita during this period, this would translate into a yearly average of between 375 and 450 crew members from United States ships alone, most of whom had been at sea for many months, with time to kill and money to spend in Paita. By the middle of the nineteenth century, one Paita neighborhood known as Maintope - 'maintop,' the platform at the overlap of the mainmast and maintopmast of a square-rigged ship - was known for its *pulperías* and brothels.[60]

Nineteenth-century travel accounts that deal with Paita are part of this conspiracy of silence, although most agree in describing the people of Paita as "open, hospitable and fully obliging to the foreigner."[61] The closest that some of them come to dealing with this issue is the discussion of the attractiveness of the women of Paita (plate 11). Reflecting his Eurocentric values, one Frenchman in 1836 noted that the more attractive women to be found in Paita were the creoles from Piura. And when the whaleship *Charles W.*

CHAPTER 2

Morgan arrived just before Lent, Second Mate James C. Osborn had liberty on Shrove Tuesday. "It was Carnaval Days with the Spanyards and the Eggs & Paint wer a flying in all Directions. about 8 oclock in the Evening I went on Shore to see the Spanish Ladys Walse And they wer so much like Home, that I Screamed."[62]

Only a few visitors make veiled allusions to the subject of prostitution, probably out of a sense of obligation to observe contemporary norms of decorum. A French naval officer, writing in 1824 about the general forwardness of the women of Chile and Peru, noted that the women of Paita and Piura were the least inhibited of all. Their movements, he said, were "voluptuous," their arms and throats were completely bare, and they "dreamed, thought and breathed of nothing but love." The same traveler described the *sange froid* of the daughters of the port captain, Panchita and Jesusa Otoya, in fending off the amorous advances of English, American, and French sailors, while noting at the same time that their orphaned cousin "trafficked publicly with her charms."[63]

When the pious Captain Butler anchored the New London whaleship *Chelsea* at Paita in 1835, he "sent the boat on shore for the liberty men they all came on board to keep sunday their being no english church on shore & the plase very disipated." And when the *Charles W. Morgan* returned to Paita in August 1843, Second Mate Osborn spent Sunday "a visiting the Ladyes, on board the ship George of Fare Haven. A good spree and no mistake." Whether he was referring to innocent flirtations or a commercial exchange is not clear. Certainly, however, numerous whaling masters were accustomed to allowing prostitutes on board their ships in order to discourage their crews from deserting ashore.[64]

Foreigners also commented on the sexual imbalance in the population of Paita, attributing the greater number of women to the fact that many of the male residents of Paita were engaged in maritime pursuits, either as crew

34

members on foreign and Peruvian ships, or as participants in the active coastal rafting trade.[65] One American consul observed that since so many men were employed as sailors on whaleships or in the coastal trade "there is a considerably larger [permanent] female population than male" in Paita.[66] Lack of economic support during the Peruvian males' long absences may have made casual prostitution a necessity for some women. Other foreign observers noted the bars, billiard parlors, and other places frequented by foreign sailors, allowing the reader to draw his or her own conclusion about what they were looking for.

Oblique references to one of the consequences of prostitution are found in the numerous accounts that describe the medicinal qualities of the water in nearby Piura, which, it was widely believed, cured syphilis, or what was often referred to as "lady fever" in nineteenth-century accounts. Stevenson reported that many Peruvians resorted to Piura for the "cure," since the Piura River ran through sarsaparilla groves and over fallen branches of the *palo santo* tree, which gave the waters medicinal qualities. He added that even in the dry season, when the river was reduced to an insignificant trickle, patients had themselves buried to their necks in the sands of the riverbed for several hours while they drank the sarsaparilla-impregated water and perspired profusely, a treatment that reportedly accelerated the cure.[67]

The serious literature on prostitution as a social and economic phenomenon in Latin America is scarce, probably because the subject was long considered taboo, and because the documentary sources - police and municipal records, public health archives, and the archives of institutions created to reform or regulate prostitution - are still untapped. Scholarly studies on prostitution in Europe and the United States during the nineteenth century, focused primarily on capitals and other major cities, emphasize efforts to repress or regulate prostitution or philanthropic attempts to reform prostitutes.[68]

CHAPTER 2

These studies offer some conceptual help in understanding how the world's oldest profession may have functioned in Paita. They pose the question of whether it was controlled by local authorities or simply tolerated; they ask to what degree it was internally organized; and they raise the issue of its economic impact on the community. Whereas official policies toward prostitution in Europe and the United States vacillated between prohibition and regulation of the institution, and sometimes emphasized the reform of its practitioners, in most of nineteenth-century Latin America there probably was no official policy, at least not until the end of the century, and then only in such highly urbanized areas as Buenos Aires or Mexico City.[69]

Scholarly studies of prostitution also help us to understand that, despite official condemnation, prostitution flourished in the nineteenth century, "possibly to a greater extent than before or since."[70] They also help us to focus on the implications of poverty for nineteenth-century women, in Latin America and elsewhere, living on the margins of society. And they help us to understand that in Paita, as elsewhere, prostitution offered one solution to the problem of economic hardship, since throughout history men have been willing to pay more for sexual access than for any other kind of female labor.[71]

Going beyond the insights we might glean from these scholarly efforts, logic again tells us that a small, relatively isolated port with a population made up primarily of working-class men and women of mixed racial background is an environment in which prostitution would be frowned on by officialdom and the Church, but nevertheless tolerated, largely for societal and economic reasons. It is also reasonable to assume that most of the women who were involved in prostitution in Paita did so on a part-time basis, as opportunity or economic necessity presented itself, and that almost all of them had other occupations or means of support, including local men to whom they might be married or with whom they had formed a consensual union.[72] Presumably, most of Paita's "fallen women" were independent of organized brothels or other forms of institutionalization, and little if any official effort was made to control or tax the activity of prostitutes, or to provide routine medical examinations and treatment designed to control the spread of venereal disease.

Nevertheless, the scholarly literature on prostitution in Latin America fails to help us understand the interpersonal dynamics of the institution in a peripheral area like Paita, which had marked ethnic and linguistic differences between the clientele and the practitioners, and where the results of these encounters - surely there were offspring - were a synthesis of the two cultures, a new kind of *mestizaje*.[73]

The fact that the baptismal records for the church of San Francisco de Paita were destroyed by fire in the 1880s makes it impossible to explore the possibility that many of these women had children by American sailors. But the Mexican experience during the occupation of 30,000 French troops in the 1860s is suggestive. Meyer and Sherman, in their masterful survey of Mexican history, posit: "Common sense defies that the generation of blue-eyed, light-skinned babies born in Mexico in the 1860s were [all] the product of French debauchery, but the subtle psysiognomic changes in those villages hosting a French garrison, or having one nearby, suggest that to the victor belong the spoils."[74]

3

Building the Bridge

T here is no clearer demonstration of the axiom that the flag follows trade than the case of the official American presence at Paita beginning in the 1830s. At some time in late 1832 or early 1833, several years prior to the beginning of the "boom" decade of New England whaling in the Pacific, the United States government decided that it would be a good idea to establish an American consulate in the northern Peruvian port. Of the 17 United States consulates that functioned in the Pacific coast seaports of Latin America - including what is today California - in the nineteenth century, Paita is the seventh oldest.[1] Only the consulates at Valparaiso (1812), Panama City, Acapulco, Lima (1823), Guayaquil (1826), and Guaymas (1832) predate Paita.

To Show the Flag

The reasons for establishing an American consulate in a primitive, faraway place like Paita were varied. In general terms, a United States presence served "to show the flag,"

39

even though - paradoxically - many American consuls in the nineteenth century complained regularly of the State Department's failure to provide a flag and a seal to distinguish their residences and offices.[2]

Most American consuls dealt with local government officials on an almost daily basis and on a variety of subjects. In broad terms, their duties included: political and economic reporting; promotion of trade between the United States and the host country; and the defense of American citizens and property within the consular district.

Examples of political and economic reporting from the consulate at Paita include: a despatch informing Washington of the arrival of Bolivian General Andrés de Santa Cruz in Lima in 1839, and the retreat to the north of the Chilean troops that had occupied the capital;[3] the forwarding to Washington of copies of an 1841 *Reglamento de Comercio* and an 1846 Peruvian decree opening the port of Tumbes to whaling vessels;[4] the relaying of news of the unsuccessful siege of Guayaquil by Ecuadorean General Juan José Flores and 300 troops in July 1852, after which four ships of Flores's fleet sought refuge at Paita and were seized by Peruvian authorities;[5] the reporting of the reaction in Paita - "implacable fury and hatred" - to the news of the Spanish bombardment of Valparaiso in early 1866, and the measure ordering Spanish residents of the port to remove themselves to Piura;[6] and relating the impact on Paita of the revolution, begun in Arequipa in 1867, of General Pedro Diez Canseco against the liberal regime of President Manuel Prado. On 5 January 1868, 150 of Diez Canseco's supporters overwhelmed the 25-man garrison at Paita, seized the customhouse and installed officials loyal to the rebel cause in all of northern Peru.[7]

An 1848 note to Secretary of State James Buchanan, offering the consul's services to agents of an American agricultural association interested in exporting alpacas from Peru in United States storeships, demonstrates the consul's role in promoting investment opportunities for American capital in Peru.[8] An 1852 report of the consul's trip to the Lobos Islands and the activities of American guano ships working there was related to English and American claims over the islands.[9]

Examples of citizen services rendered on behalf of expatriate Americans include an official protest to local authorities over the extortion of a bribe from a Massachusetts man who operated a cotton plantation near Paita in 1839, and the printing and distribution of broadsides to inform American whalers of the outbreak of hostilities between the United States and Mexico in 1846, warning them not to call at Mexican ports.[10]

Serving the Whaling Industry

The United States consuls that served in Paita from 1833 to 1874 did all of these things, but for the purposes of this study two aspects of their work merit special attention. They are: the regular consular despatches sent to Washington, providing detailed information on United States whalers and merchant vessels that called at Paita; and the defense of American seamen and officers who found themselves in the Peruvian port. I will examine the information found in ship reports in chapter 5. At this point, however, it seems appropriate to examine the consul's role as a protector of sick and destitute American sailors and those who had run afoul of their shipmasters and ended up in Paita.

For many a New England lad, shipping as a crew member aboard a New Bedford or Nantucket whaler for a "cruise" of two or three years in the Pacific promised adventure and a handsome reward when the ship's oil was sold and shared out among crew members and officers.[11] But it also meant long hours of hard work in dirty, crowded ships; months at sea with often inadequate food; the chance of serious injury, illness, or death in a place far from home;

CHAPTER 3

and the possibility of working under a tyrannical master. Because of the hardships and uncertainties involved, skilled seamen generally avoided service on whalers.[12]

Of all the calamities that could befall a Yankee sailor far from home, the one that might first leap to the mind of many - being tossed in the local jail - appears to have been a rare phenomenon in the annals of the American consulate at Paita. Thomas Roe mentioned in his 1831 journal that two members of the *Chelsea*'s crew tried to desert in Paita. They were seized, robbed, and jailed by local authorities. Their captain sent them food, which was allegedly taken by their jailers, and they nearly starved before the master ransomed them for $15 each.[13] Or, perhaps, landing in jail was so common a phenomenon that it did not rate special comment in the regular consular despatches. In any case, during the 42-year existence of the United States consulate at Paita only one case of an imprisoned American seaman is documented, and that case involved a complaint against the consul!

In December of 1856 an American seaman named William Rose wrote to the secretary of state to complain of the treatment that he had received a year earlier at the hands of the municipal authorities in Paita and the American consul. He and three other sailors had been imprisoned for some unspecified misdemeanor. When offered their freedom if they would agree to ship out immediately on an American whaler in port at the time, Rose refused, allegedly because of the bad reputation of the ship's master. The consul opted to keep him in jail, refusing even to allow him to go to the hospital.

One interesting aspect of this incident is Rose's description of "the worst of holes called a jail" in Paita. He was confined for 36 days in a bamboo and mud common cell measuring 20 by 28 feet, with a damp dirt floor on which he slept, or would have slept, were it not for the numerous fleas and "vermin of all discripion [*sic*]" that made it impossible. The cell's only light came in through a 4-by-5-inch opening in the door. Rose was not allowed a change of clothing during the entire time of his confinement, and he had to purchase his meager rations from the jailer for 18 3/4 cents a day. Fortunately, some mates in port at the time supplemented his diet with bread and other items.[14]

Rose's account of his sojourn in the Paita jail does not mention the circumstances of his release, but it is most likely that he eventually shipped aboard an American whaler that called at Paita. Rose's complaint is also interesting because it highlights the dilemma faced by the consul, who was by law responsible for the cost of housing and feeding destitute, ill, or miscreant Americans during their stay in port, not a small responsibility as we shall see.

Another of the consul's duties involved trying to locate underage boys who had gone to sea without parental permission, or seamen whose families had lost track of them and were worried about their safety. One 16-year-old lad named John J. Connell had shipped aboard the *Anaconda* of New Bedford as a carpenter, without his parents' permission, using an alias and lying about his age. His family engaged a New York law firm to find him and posted the money to pay for his return passage, but the master of the *Anaconda* refused to release him. The consul was asked to look out for the boy and intervene on his family's behalf.[15]

Given the slowness of communications, sometimes the "disappearance" of an American seaman was resolved even before the consul was able to inquire or intervene on his behalf. Such was the case of the brothers Orlando and Elisha Carr, who had shipped aboard the *Ebenezer Dodge* without the permission of their father. By the time the consul in Paita was alerted to their disappearance, one had already departed for Nantucket via Cape Horn, and the other was living at Talcahuano, in southern Chile.[16] In other instances, the object of the consul's search never did surface, as in the case of one George Harris of Brooklyn, whose brother wrote to the State Department in May of 1854 asking for assistance in finding him. The only information that the consul in Paita was able to uncover was that Harris had worked for two years for a Mr. Taylor in Arica, in southern Peru, and had disappeared into the wilds of Bolivia.[17]

The most bizarre of these welfare cases involved a young sailor who shipped aboard the *Christopher Mitchell* of Nantucket in December of 1848 bound for the Pacific. Six months later, off Paita, the sailor became ill and the master discovered that "he" was a young woman named Ann Johnson. The captain immediately headed for port, in order to deliver her into the hands of the consul. Since it was apparent that she would be unable to ship out again on another whaler, the consul refused to receive her until the captain gave him a draft for $300 drawn on the ship's owners in Nantucket. They later protested to the State Department about the consul's arbitrary actions and tried to have him dismissed. What seemed most to bother C. Mitchell and Company of Nantucket was the $236.25 that the consul spent on clothing, board and laundry, incidental expenses, and cabin passage back to the United States for the hapless female whaler.[18]

The consul at Paita also had his share of accidents and disasters to deal with. In early June of 1842 the *Orbit* of Nantucket went aground at a place called Punta Roncadora. The master and crew of 27 escaped with the clothes on their backs, and five days later they arrived at Paita completely destitute. An English carpenter who had shipped in Valparaiso was the only casualty. The consul subsequently returned to the site of the wreck with the master in order to salvage the ship's cargo, sails, rigging, timbers, and other materials. He later was named agent to sell the effects of the *Orbit* at auction. Part of the proceeds of the sale went to reimburse him for the $1,089 that he had advanced the *Orbit*'s crew for clothing and room and board. By the end of June some crew members had shipped aboard other whalers, while others were still under the consul's care.[19]

Another ship, the *Ann Alexander* of New Bedford, was struck in the bow by a whale, foundered, and sunk in August of 1851. Nearly a month later the crew was put ashore at Paita by another ship, and by mid-October the consul had spent $270.50 of United States government funds in assistance to the destitute crew members, all of whom by then had either shipped aboard another whaler or returned directly to the United States.[20]

Similar incidents that demanded the consul's intervention occurred with depressing regularity. The barque *Emma* burned to the waterline in Paita on 14 October 1852. Arson was suspected. The *Brandt* of New Jersey sank off Chatham (San Cristobal) Island in the Galapagos in December of 1853. The destitute crew was brought to Paita aboard a Peruvian brig. The *Venezuela* of San Francisco was lost on a reef off Ecuador in December 1853. The loss was certified by the consul in Paita by submitting one-half of the ship's registry to Washington. A year later the *Memnon* of Nantucket burned to the waterline in Paita bay, set afire by some of the crew. The men escaped with only the clothes on their backs. In June of 1857 the *Susan* of New Bedford was lost off the Esmeraldas coast in Ecuador. The captain and crew arrived in Paita 20 days later. The crew of the *William and Eliza* tried to burn her off Tumbes in April of 1857 while the master was ashore buying supplies. The American vice consul in Tumbes advised the captain to make for Paita, since the authorities in Tumbes were deemed unlikely to cooperate in trying the plotters and because the port had no jail.[21]

Murder and mayhem on the high seas also intruded on the lives of the American consuls in Paita. In late August of 1839 an oil barrel shattered the leg of Captain Morselander of the whaler *Charles* of New Bedford. Nine days later, in Paita, the gangrenous limb was amputated, and five days after that Captain Morselander died in the house of his friend and agent P. Hynes.[22] In August of 1844 a black crew member of the *Ontario* of Nantucket named

CHAPTER 3

William H. Corsa murdered first mate Andrew B. Brooks off the Atacama coast. Corsa escaped overboard but was later apprehended by Peruvian authorities. The captain of the *Ontario* asked the consul in Paita for help in having Corsa remitted to the United States in irons to stand trial.[23] In late July of 1842 third mate Lawrence Marshall of the *Orbit* was stabbed to death by a Chilean crewman. He was buried at sea at United States government expense. The Chilean escaped. In late 1852 a boatsteerer of the *Andrews* stabbed Captain James Nye in the neck "in self defense." The boatsteerer was beaten, put in irons in the ship's blubber room, and later tossed overboard at night in heavy seas off Charles Island in the Galapagos. The consul reported him drowned.

Sometimes, as in the case of the boatsteerer of the *Andrews,* sailors reacted violently against abusive masters. In other cases, as seen earlier, crew members sought revenge by burning the ship. In the most egregious incidents of abuse, the consul at Paita felt obliged to report the offenders to authorities in the United States. Such was the case in 1854 when the consul denounced Captain Henry D. Norton of the New Bedford whaleship *Hector* to the office of the United States Attorney for the District of Massachusetts, in Boston, for his "most shocking and cruel" abuse of his crew. The United States Attorney acknowledged receipt of the consul's complaint and his "evidence," but advised that an answer would be delayed. There is no indication of how the complaint was resolved, or of the exact nature of the charges against Captain Norton. In another case reported by the American consul at Lambayeque in 1866, a boy named Charles Thompson was repeatedly hung by his thumbs and beaten by the shipmaster, to the point where he almost died and had to be sent to Panama via steamer by the consul.[24]

The preeminent historian of the New England whaling industry, Alexander Starbuck, acknowledged that some crew members may have been forced to desert their ships in order to escape from the impositions and severities of brutal captains, but insists that "such cases were undoubtedly very rare."[25] Evidence derived from the consular despatches from Paita suggests, however, that all manner of abuses were committed regularly by the Yankee masters.

A variety of scams concocted by American shipowners, captains, and merchants to cheat greenhands became standard practice as the whaling industry reached its apogee. Most of the abuses reported by the consuls in Paita charged with the protection of American seamen involved shipmasters trying to elude their financial responsibilities toward crew members. From the early years of the whaling boom, Yankee masters were in the habit of anchoring in the bay to take on water and supplies without formally calling at the port or depositing their papers with the consulate, for which a small fee was paid. This was a violation of an 1813 United States law. Many captains shipped and discharged crew members without the consul's intervention, thus also avoiding the payment of a fee. The laxness of the Peruvian port authorities at Paita facilitated this kind of abuse. Since by Peruvian law whaling ships were exempt from port fees, the authorities were very tolerant of this practice because they stood to gain nothing by enforcing regulations about ship calls.[26]

The case of seaman Charles L. Hatch of the New Bedford bark *Pacific* (2), Joseph L. Smith master, was typical of this kind of abuse. While anchored off Paita, Captain Smith coerced Hatch to leave the ship and to relinquish payment of the "lay," or share of the ship's oil cargo to which he was by law entitled. At the next port of call Smith filed desertion papers on Hatch, pocketing his share. Some victims of this practice were able to ship out immediately on another whaler. Others, in part because of the seasonal nature of whaling, were unable to find a new ship. From mid-December to the end of August fewer whalers called

at Paita than during the rest of the year.[27] Still other seamen were too ill to find work, joining those who became
government charges.[28]

One consul wrote eloquently on the arbitrariness
of some shipmasters in a report to the secretary of state in
early 1863: "I am surprised often at the palpable and willful evasion of the law by many shipmasters, who fancy
themselves to possess...a roving commission to ship men
at Islands and ports...where there is no consul, upon such
selfish terms that seamen being furnished with a little clothing at enormous prices are held peons or slaves to work
out long services without adequate compensation and most
frequently with none at all."[29]

Ever the apologist for the industry, Starbuck, in describing the reasons for its decline, obliquely addressed the
issue of dishonest captains. In placing part of the blame for
the collapse of whaling on the "exhorbitant charges and
percentages" demanded by United States consuls, he indirectly exonerated the masters for illegally discharging seamen to avoid paying consular fees. Starbuck added: "In
many cases justice...seems to have been meted more in accordance with the requirements of the income of our representatives than with those of abstract right."[30] The fact
that several of the men nominated to serve as United States
consul in Paita declined the honor or resigned shortly after
their arrival in Peru because they found the remuneration
to be uncommensurate with the cost of living suggests that
a consul's income was meager, despite the fees that he was
authorized - and obliged by law - to collect from American
ships that called in his jurisdiction. Clearly then, with regard to the matter of fees collected by the consul, "justice"
was in the eye of the beholder.

BUILDING THE BRIDGE

The American Hospital in Paita

The consul's dealings with legitimately sick and destitute seamen, many of whom died and were buried in Paita, is yet another chapter of this story. It involves the personal tragedies of anonymous Americans who died far from home and family, and it involves an institution about which little or nothing is known, the American Consular Hospital. A few names will help make the tragedy of those sailors who lost the final battle against the Leviathan seem more personal. Adam Hazard, the African-American boatsteerer of the brig *Fabius* of Nantucket, was discharged before the American consul in Paita in accordance with the law on 21 November 1840 and died the same day. His effects, of little value, were given to a mate who was with him when he died. Joseph Ellis of the Nantucket ship *President* was landed sick in June of 1841 and remained on shore until he died on 28 October, leaving no personal effects. The black cook of the ship *Rousseau* of New Bedford was put ashore in Paita in September 1841, "deranged" from falling down a hatchway. He died on 29 October. Lot Luce was put ashore from the *Joseph Maxwell* of Fairhaven in February 1842 and died on 10 May. James Hodges of the Nantucket ship *Aurora* was left in Paita in March 1842 and died on 16 June. Holmes Jernegan of the New Bedford ship *Charles W. Morgan* was sent ashore on 2 February 1844 and died on 7 March. The consul forwarded his sea chest containing clothing and bedding to his next of kin. Henry Bardley, the cook of the ship *Corinthian* of New Bedford, was discharged before the consul on 3 June 1844 and died a week later. Amasa M. Webster of the Fairhaven bark *Hesper* died at Paita on 13 September 1845.[31]

Most of the American seamen who died at Paita in the 1840s and 1850s were taken in longboats to be buried in the cemetery for foreign seamen located at a place called Cabo Blanquillo.[32] Even if it were no more appalling than the municipal cemetery at Paita, described by an American traveler in 1861, Cabo Blanquillo must have been a dreadful place. Baxley depicted the Paita cemetery as being enclosed by a bamboo and mud daub fence, with wooden grave markers instead of marble monuments and epitaphs. Some of the graves had been disinterred by the constant wind, and in the center there was a pyramid mound formed from hundreds of sun-bleached skulls, surmounted by a wooden cross.[33]

Of course, not all of the Yankee whalemen treated in the hospital in Paita ended up at Cabo Blanquillo. But no matter whether they went to their final reward in Paita, or survived to ship out on another whaler, the sojourns of these American seamen in Paita cost the United States government money. Unfortunately, the consular records for this period are not complete enough to form a comprehensive image of this facet of the consul's duties for the period 1832-74 or the costs incurred. Nor do they distinguish between sick sailors who were unable to work, and destitute seamen who did not have funds for their passage home and who were unable or unwilling to reship aboard a whaler. We know that in the first half of 1835 the acting American consul in Paita spent $1,444.53 on the care of sick and destitute American seamen, and that the consul general in Lima refused to reimburse him.[34] In 1837 the same official reported that the Peruvian coast was "infested" with runaway foreign seamen, primarily from whalers.[35]

During the period between 1 July 1839 and 31 March 1854 United States consuls in Paita submitted to Washington vouchers for the care of sick and destitute American seamen totalling at least $47,191. This expenditure of United States government funds averaged $266 a month over this 15-year period.[36] Although some of the vouchers list the names of the beneficiaries of consular assistance, it is not possible to even estimate how many American sailors were aided by the consulate during this period. If, however, the $266 average monthly expenditure for maintaining sick and destitute seamen in the hospital or a local boardinghouse is divided by the daily rate of $1.50 that was in effect for at least part of this period,[37] it would appear that an average of about 177 man/days of hospital care or room and board were provided each month by the consul in Paita over that 15-year period. Eventually the auditors in Washington began to question this practice.

The consular records for the early years do not indicate the names or nationalities of the physicians who treated American seamen in Paita, or when professional medical attention became available for the first time there. It seems logical to assume that there were Peruvian doctors practicing at the Bethlemite hospital in Piura early on, but until the beginning of the whaling boom in the 1830s there was little in Paita to attract a medical professional. One 1812 account of a Nantucket captain who was struck in the head by a whale's jaw off Paita reveals that medical help - in the person of a 69-year-old surgeon - had to be brought from Puna Island, in the Gulf of Guayaquil, to minister to the injured whaleman. The Yankee captain remained at Paita almost three months under the care of the "skilled" Peruvian doctor, and lived to tell the tale.[38]

While the importance of the role played by the consular hospital in Paita is abundantly clear, the institutional history of this hospital is shrouded in mystery.[39] It appears that it was created during the tenure of the first American consul to serve in that port. The first mention of the "consular hospital" in Paita is found in a voucher for reimbursement of monies spent on behalf of ill and destitute seamen dated 30 June 1849. Another voucher dated 30 June 1851 refers to "the attending physician of the consular hospital." An even more explicit reference is found in a consular despatch dated 12 September 1852 regarding a sick seaman from the whaler *Louisiana* of New Bedford, whose master left him in Paita without making financial arrangements for his care. The despatch includes a certificate signed by the "resident physician of the Am[erican] Hospital of this place," A Dr. Fayette M. Ringgold; the ex-physician of said hospital, a Dr. Thomas J. Oakford, at the time United States consul in Tumbes; and a Dr. T.F. Churchill, "an English M.D. casually resident in this place."[40]

The genesis of the "Am[erican] Hospital" (the consul's capitalization) and the date of its organization are unknown. None of the consular records indicate exactly where in Paita it functioned, its size, or how it was staffed except to note that at the beginning North American doctors were in charge. Indirect evidence suggests that the first resident physician was a Dr. Thomas J. Oakford. Sometime around October of 1852 Oakford was replaced by a Maryland physician, Dr. Fayette M. Ringgold, who had been American consul in Arica for two years.[41] Ringgold, who later became consul in Paita, informed the secretary of state in 1854 that "no other hospital exists in Paita but that established by the American and English consuls." He explained that seamen were admitted by their respective consuls upon showing proof of citizenship, and that "the attendance has always been good and the accommodations excellent."[42]

Seamen admitted to the American Hospital were treated for a variety of maladies associated with whaling, several of which were fatal. According to one contemporary observer, an American doctor, many suffered from scurvy, gastritis, and liver and kidney disease owing to the shipboard diet of salted beef, salt pork, and hard biscuits. Others were treated for tuberculosis and "dropsy [edema] of the chest or abdomen" as a result of long exposure to sea air and water, and to the dank environment of their quarters in a whaleship forecastle. Still others came to the hospital with fractures resulting from shipboard accidents, or wounds received in altercations on the high seas.

In addition, "the habits of whalemen when in port" resulted in a high incidence of venereal disease, which shipmasters generally treated at sea with doses of mercury that provided immediate - if very dangerous - relief, leaving many "wracked with neuralgia, rheumatism, carious bones, ulcers horrid beyond description, and a class of disease which takes a long time and a great deal of care to eradicate." In contrast to the preferred shipboard treatment for venereal disease, in Peru the affliction was treated with doses of sarsaparilla, guaiacum, and sassafras. In time, as suggested by the account books of whaleships that called at Paita, the masters of American ships also began to buy and use sarsaparilla bark or root, as well as tincture of guaiacum, to treat venereal disease. Often, sick or injured crewmen remained at sea for months before their ship's routine call at port, exacerbating their condition and even threatening their lives. Many seamen deserted their ships when the masters refused to accept that they were too sick to continue the cruise.[43]

Like most human institutions, the American Hospital at Paita had its ups and downs. In late 1861 the acting consul reported that the hospital, which at the time housed 18 seamen - many of whom were in a "very precarious state of health" - was threatened with closure. Two foreign-owned trading companies in Paita had threatened to cut off advances because the State Department auditors had refused to honor drafts issued by the consul for nine months' worth of hospital expenses totaling $8,451. The government auditors had ordered a suspension of payments because of perceived irregularities in the consul's accounts. His temporary absence from post made it impossible to resolve the question to the satisfaction of the consulate's creditors.[44]

Nearly a year later the new appointee, who replaced the consul whose accounting had been questioned by government auditors, informed the State Department that the closing of the American Hospital some nine months earlier had been prejudicial to sick American seamen. He said they had to rely on private charity, or beg and borrow the money for their return passage to the United States. The alternatives were to become wards of the Peruvian government, to enter the newly organized local hospital, to ship out while still ill, or to wander the streets of Paita "sick...and in tatters." The consul informed the Secretary of State that he had lodged two sick sailors in private boardinghouses at a cost of $1.50 a day, plus medical treatment.[45]

In a subsequent despatch, the consul - who also happened to be a physician - informed Washington that he had reopened the American Hospital in Paita. He argued that the cost - $1.50 per day per patient - would be less than lodging sick sailors in boardinghouses, which he described as "unventilated, unlighted, smokey and filthy hovels" where the care was inadequate. The consul concluded that the failure of the United States government to provide adequate medical care for men engaged in "one of the most important branches of our commercial industry [was] a reproach upon the nation." In order to finance the revived hospital, he increased the three-month payment required of shipmasters who wished to legally discharge a sick seaman from $30 to $36, which he said was "willingly" paid by most masters.[46]

Appended to these two consular despatches are two documents actually generated by the American Hospital in Paita. Dated 30 October 1862 and 31 December 1862, they list the seamen admitted to the hospital, the date of their discharge, and the cost of their care, which was reimbursed by the consul. The two reports list 37 seamen treated during the last half of 1862, a total of 1,678 man/days, by a Dr. Domingo Davini. They do not list the nature of their illnesses. A similar voucher dated 31 March 1863 lists 27 American seamen admitted to the American Hospital in Paita during the first quarter of 1863, of which all but nine had recovered enough to be discharged. The average length of hospitalization was 44 days, but five men on the list had been hospitalized during the entire quarter. The cost of hospitalization was $1,764.[47]

The creation of an American-run hospital in Paita, financed in part by the United States government and devoted to the care of American seamen, was a novel but practical solution to a serious problem. As to its success, the fact that almost all of the deaths of seamen recorded in the consular despatches from Paita seem to predate the establishment of the American Consular Hospital some time in the mid-to-late-1840s, indirectly suggests a high recovery rate for patients of that institution. As consul Winslow suggesed, sending a sick sailor to the municipal hospital in Paita, or even to the one in the departmental capital Piura, was not feasible.[48] The standards of hygiene and professional care in those local institutions were probably far below the modest levels attained by an expatriate American hospital in the mid-nineteenth century. Nor was it practical to return the ill seaman to the United States by sailing vessel around Cape Horn. But the idea of a seaman's hospital was not in itself unique in the Peruvian context. In 1803 Stevenson visited the Espiritu Santo seaman's hospital located at Bellavista, between Callao and Lima. He noted that a small amount was deducted from the pay of every sailor who entered the port to support the hospital.[49]

It is not within the scope of the study to document the existence of American-run and government-financed seaman's hospitals at other Pacific whaling ports in the nineteenth century. However, there appears to have been an American hospital at Tumbes in the 1850s, after that port was officially opened to American whalers.[50] There was also an American Hospital in Callao, whose owner and resident physician was Dr. Charles Winslow, the third American consul in Paita. Prior to coming to Peru, Dr. Winslow was employed as a physician and surgeon at the seamen's hospital established at the Hawaiian whaling port of Lahaina in the 1840s.[51]

Moreover, the visit of Dr. Henry Willis Baxley to Pacific ports in South America, Hawaii, and California in 1860-61, which resulted in the work frequently cited in this study, was related to the functioning of consular hospitals. In 1860 President James Buchanan named Baxley, a professor of medicine at the University of Maryland, as special

55

commissioner of the United States government to study and reform hospital conditions in United States consular districts in the Pacific.[52] In his book *What I Saw on the West Coast of South and North America and at the Hawaiian Islands* Baxley does not specify that his mission was to study American-run, United States government-financed consular hospitals. Nor does he mention the American Hospital in Paita in his description of that port, perhaps because his visit coincided with the period during which the hospital was closed. In any case, as we shall see in chapter 8, with the dramatic collapse of the whaling industry in the mid-1860s, the sharp decline in the number of United States ships calling at Paita, and the increasing economic demands on the United States government because of the Civil War, the rationale for the maintenance of the American Hospital at Paita - indeed for the existence of a United States consulate in that port - disappeared.

4

The First American Consul in Paita

In March of 1833 a gentleman from Auburn, New York, who had been nominated to serve as American consul in Paita, wrote to Secretary of State Edward Livingston asking if the United States government would pay his passage to Peru, and inquiring about the salary for the proposed appointment. Two weeks later, having received a negative answer to his first question, he informed Livingston that he would try to find passage on a whaler bound for the Pacific. That apparently fell through, and in early June he declined the appointment due to "family circumstances."[1]

In January of 1835 a gentleman from Philadelphia who had just returned from Peru wrote to Secretary of State John Forsyth regarding his son's candidacy for the position of American consul in Paita. At the end of the month, in another letter, he implied that his son had already departed for the Pacific, but the consular records in the National Archives bear no witness to his ever having served in Paita.[2]

At exactly the same time, in a note to Secretary of State Forsyth, a British merchant resident in Paita acknowledged receipt of his appointment by President Andrew Jackson and his commission as acting American consul in Paita, and informed the Secretary that he had entered into duty.[3] Mr. Charles Higginson, who had resided for many years in South America, was strongly supported for the position at Paita by the Massachusetts whaling industry. A petition signed in February 1834 by 103 shipowners and merchants from Nantucket, New Bedford, and Fairhaven urged the secretary of state to appoint Higginson. The whalemen complained that Paita and other Pacific ports had become "infested with a set of idle and vicious foreigners, mostly deserters from whaling ships; always ready on the arrival of ships, to entice the crews to pilfer and desert."[4] Obviously, they expected that the presence of a diplomatic representative of the United States in Paita would somehow help remedy this situation.

Higginson, who was also the British consul in Paita, represented the interests of the United States from January of 1835 until mid-1839. In July of 1839 a young merchant from New York City, Alexander Ruden, Jr., formally took possession of the consulate.[5]

The Expatriate Merchant Prince

With the appointment of Ruden, who had lived in Paita earlier, the New England whalers and other Americans who called at Paita gained a defender of their interests. At the same time, regular consular despatches full of useful information began to arrive from Paita. In introducing the first American consul to serve in Paita and his successors, as well as other expatriates from the North Atlantic world who

THE FIRST AMERICAN CONSUL IN PAITA

for a variety of reasons ended up in Peru, I hope to answer two questions. The first: who was Alexander Ruden, Jr., and how did he become involved in trade on the West Coast of South America in the late 1830s, a pursuit that led him to seek the position of United States Consul in Paita? The second: what relevance does Ruden's story and that of the others that will be presented in this study have to the story of Paita's role as a socio-economic bridge between Peru and the United States in the early nineteenth century?

The answer to the first question is incomplete, but using the research techniques of the genealogist a general outline of his background can be pieced together from United States decennial census records, *Longworth's New York Directory* for the period 1807-43, *Doggett's The New-York City and Co-Partnership Directory* for 1843 on, the *Index to Wills for New York County, 1662-1850*, and United States consular records. Unfortunately, there are no letters of recommendation on Ruden's behalf or other communications relating to his appointment as consul in Paita in the diplomatic records held by the National Archives, which is often the case for American consuls in the nineteenth century, and which generally provide useful background information. The answer to the broader question about the relevance of biographical information on expatriates working and living in Peru in the nineteenth century will, I hope, become evident.

The Rudens were New York merchants of modest means whose origins are unclear. The tantalizing but elusive data from the above-mentioned sources suggest that Alexander Jr. was the grandson of merchant Jacques Ruden, who apparently arrived in New York around the year 1800 and died in 1808. The surname may be of Scandinavian origin, with a possible French connection. Alexander Jr.'s brothers were named Jacques and Emmanuel. Alexander Sr. apparently emigrated from Europe with his family and became a merchant after the death of his father.

CHAPTER 4

In the 1820 census there were nine persons in Alexander Sr.'s household, including a female African-American slave. In addition to Alexander Sr. and his wife Magdaline, both of whom were over 45 years of age, the household contained an older woman - probably Alexander's mother Rachel - and five children ranging in age from under ten to 26. His name suggests that Alexander Jr. was probably the first born surviving male, which would place him in the census' 16-26 age category. If his father's family arrived in New York directly from Europe around 1800, Alexander Jr., like his father, may have been born in Europe.

Nor do we know what merchandise his father sold in New York, or how successful he was. Indirect evidence - the fact that his places of business were located primarily on Canal, Front, South, and Pearl Streets - suggests that he dealt in ship stores. We do know, however, that in the 25 years between 1808 and 1832 the senior Ruden catered to the trade at 16 different New York city addresses, which suggests that he was operating out of rented quarters and that he was only moderately successful as a merchant. The fact that the family owned a household slave in 1820 does not necessarily make them wealthy.

In the year 1832, however, Alexander Sr. set up shop at 196 Canal Street, where he remained through the early 1840s. The fact that the senior Ruden did business for at least 13 years at the same address points to ownership of his own building. Four years prior to the move to Canal Street, in the directory for 1827-28, he began listing his occupation as "broker" - no specifics - which also suggests an upwardly mobile career pattern. The family's residential pattern - they had at least 19 addresses between 1801 and 1836 - suggests that they were renters and tends to confirm their relatively modest means. Alexander Sr. died in 1847 or 1848, and his will was probated in New York on 18 July 1848.[6]

THE FIRST AMERICAN CONSUL IN PAITA
On His Own

Alexander Jr. most likely worked with his father in trade in his early years, just as his father had done before him. It is probably no coincidence that the junior Ruden first appeared in Longworth's *Directory* as an independent merchant - with a shop just a few doors up the street from his father's business address at 48 Wall Street - the same year that Alexander Sr. began listing himself as a broker. Two years later (1830-31) Longworth's listed Alexander Jr. as a "shipmaster" with headquarters at 38 Dominick Street, but by the following year (1831-32) he was again a merchant. As one might expect, during the initial years as a fledgling businessman in New York City, Alexander Jr. lived with the senior Rudens, just as his father had done before him. At some time before the publication of the 1832-33 *Directory* the young merchant-shipmaster, who was probably in his early 30s, apparently sailed for Latin America to seek his fortune, since his name does not appear in the *Directory* for that year or for any subsequent years of my search.

Off to seek his fortune in Latin America in 1832 or 1833, Alexander Ruden, Jr., probably did not have a specific destination or career goal in mind, other than to build upon his mercantile experience and perhaps his association with his father in order to make his mark in the world. The stories brought back from Latin America by American merchants and whalers who had penetrated the initially very lucrative market of those newly independent republics were probably a powerful stimulus to this ambition.

But by the mid-1830s the Latin American market was not the virgin territory that British and American traders had encountered a decade or two earlier, when the mercantile system that bound the Spanish empire and - at least in theory - excluded foreign traders began to disintegrate. By the time Alexander Jr. arrived on the scene the bloom had faded. Latin American markets had become saturated with the flood of European manufactured goods valued at between four and five million dollars. At the same time, the more mundane products exported by the United States - flour, lard, and coarse cloth - posed a direct threat to the hegemony of Latin American producers and merchants, especially in Peru. By the 1830s, after the failure of the free-trade liberals of the previous decade to create an open economy, the Peruvian market became closed and xeno-phobic. In the words of a British envoy to Lima in 1838, "jealousy and hatred of foreigners are undoubtedly the two most powerful characteristic passions of the Peruvian na-tion." Moreover, the anticipated economic revival associ-ated with political independence did not materialize, in part because of the extensive damage to infrastructure, dispersal of labor, and the significant capital loss caused by the 15-year independence struggle. Nor did the influx of foreign capital to Latin America, in the form of loans and direct investments, live to expectations after the 1825 crash of the London bond market.[7]

All this leads me to assume that young Alexander found it difficult to establish himself in the Latin American trade. He may have initially tested the market in the Carib-bean. The fact that his younger brother Jacques was a cigar merchant suggests contacts with Cuba. He may have moved on later to Brazil, Buenos Aires, or Valparaiso prior to ar-riving in northern Peru in 1836 or 1837. In June of 1838 Ruden was back in Valparaiso, from whence he wrote to the secretary of state announcing that - on the presump-tion that his appointment to the consulate in Paita had been confirmed - he planned to return to Paita soon.[8]

For reasons that are not clear, however, Ruden re-mained in Lima until the middle of 1839, after the collapse of Andrés de Santa Cruz's Peru-Bolivian Confederation. This was a time when, in Gootenberg's words, "North American commercial and political stock had plummeted in Peru."[9] Ruden's prolonged sojourn in Lima suggests, however, that he may have found a measure of commercial success in the Peruvian capital, despite the hostile trade environment.[10] Then, as we have seen, on 1 July 1839 he formally took possession of the consulate, which he probably saw as a buffer against the uncertainties of the times. But by this time his contacts with his family in New York appear to have become attenuated.[11]

Alexander Ruden arrived in Paita at a difficult time for foreign merchants. After the expulsion of the Bolivarian, free-trade liberals, the new republic's landed, commercial and military-bureaucratic elites adopted a stance of defiant economic nationalism, directly challenging the trend that, well into the 1840s and the beginning of the guano boom, had favored accommodation with the liberal world economy.[12] Despite these odds, however, Alexander Ruden appears to have been very successful as consul and merchant in Paita.

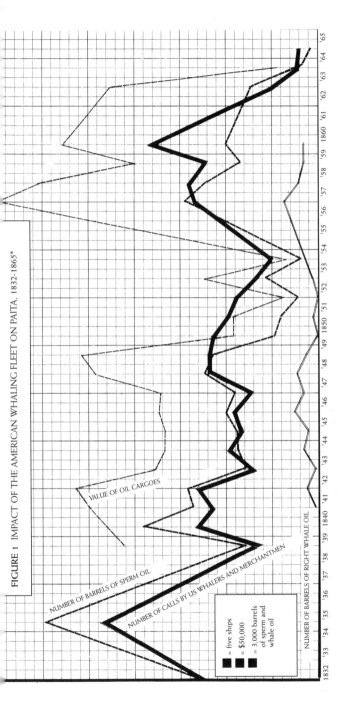

FIGURE 1 IMPACT OF THE AMERICAN WHALING FLEET ON PAITA, 1832-1865*

NUMBER OF BARRELS OF SPERM OIL

NUMBER OF CALLS BY US WHALERS AND MERCHANTMEN

VALUE OF OIL CARGOES

NUMBER OF BARRELS OF RIGHT WHALE OIL

■ = five ships
■ = $50,000
■ = 3,000 barrels of sperm and whale oil

• Sources: NA, T600, Consular Returns for Paita for the period 1832-65. Note: Returns for the years 1833, 1834, 1836, 1837 1838, 1855, 1856, 1861, and 1862 are missing, making it necessary to bridge those gaps in the graph. Data for the years 1835, 1839, 1854, 1857, 1858, and 1859 is incomplete, making it necessary to reconstruct figures for the lacunae on the basis of available data, taking into account the seasonal nature of ship calls at Paita as demonstrated in existing data.

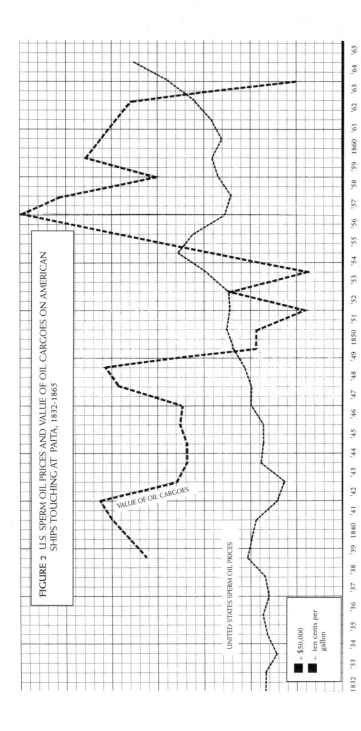

FIGURE 2 U.S. SPERM OIL PRICES AND VALUE OF OIL CARGOES ON AMERICAN
SHIPS TOUCHING AT PAITA, 1832-1865

VALUE OF OIL CARGOES

UNITED STATES SPERM OIL PRICES

■ = $50,000
■ = ten cents per gallon

• Sources: NA, T600, Consular Returns for Paita for the period 1841-65; Starbuck, *History of the American Whale Fishery*, 660.

Map 1. Paita and its hinterland. (Map by Stephen D. Oltmann)

Map 2. The coast of Peru and Ecuador, showing the river systems and the location of Paita. (Map by Stephen D. Oltmann)

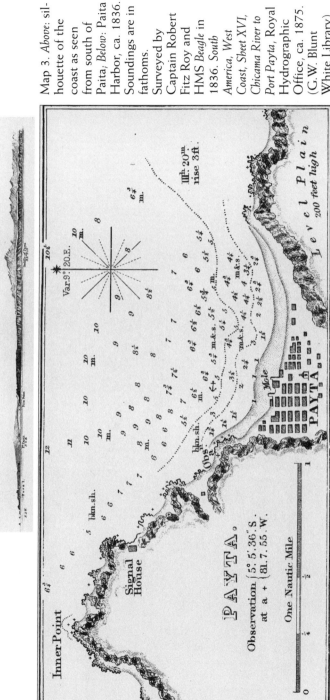

Map 3. *Above:* silhouette of the coast as seen from south of Paita, *Below:* Paita Harbor, ca. 1836. Soundings are in fathoms. Surveyed by Captain Robert Fitz Roy and HMS *Beagle* in 1836. *South America, West Coast, Sheet XVI, Chicama River to Port Payta,* Royal Hydrographic Office, ca. 1875. (G.W. Blunt White Library)

Plate 1. This street view illustrates the one- and two-story dwellings of Paita, many of them sided with split bamboo, and most thatched with palm fronds. Drawn by Victor Adam, lithograph by Bichebois, in A. de la Salle, *Voyage autour de monde...sur la corvette* La Bonite, *album historique*, Paris, 1851. (Courtesy the George Peabody Library of The Johns Hopkins University)

Plate 2. Located at the eastern edge of Paita were the Convent and Church of La Merced, the moral and spiritual center of Paita. Drawn by Bayot, lithograph by Mayer, in A. de la Salle, *Voyage autour de monde...sur la corvette La Bonite, album historique,* Paris, 1851. (Courtesy the George Peabody Library of The Johns Hopkins University)

Plate 3. In 1741, Lord George Anson, with a British fleet of five ships, sacked and burned Paita. The port was similarly destroyed in 1557, three times in the 1600s, in 1720, and again in 1819. Engraving by Muller, in Richard Walters, *A Voyage Round the World...*, London, 1748.

Plate 4. In Peru, the native hammock and the Spanish guitar came together in a way that was particularly seductive to New England mariners. Lithograph, Lima, ca. 1860. (G.W. Blunt White Library)

Plate 5. The great majority of whaleships calling at Paita came from the Massachusetts whaling ports of New Bedford, shown here, and Fairhaven (out of sight at right, facing New Bedford across the Acushnet River). The world's principal whaling port in the mid-nineteenth century, New Bedford was the capital of the whaling empire in which Paita served as outpost. Drawn by J.W. Barber, engraved by S.E. Brown, ca. 1839. (Courtesy The Whaling Museum, New Bedford, Massachusetts)

Plate 6. In front of the Church of San Francisco de Paita, the public market was the economic center of Paita. The large timbers in the right foreground appear to be salvaged ship timbers, reflecting the port's role as a supply and repair point. Lithograph by Bayot, in A. de la Salle, *Voyage autour de monde...sur la corvette* La Bonite, *album historique,* Paris, 1851. (Courtesy the George Peabody Library of The Johns Hopkins University)

Plate 7. A view along the waterfront of Paita suggests the rugged, desolate landscape in this arid region. Yet, as this view indicates, trade had brought enough prosperity to Paita for multistory houses, many with plaster facades and galleries, to line the streets by the late 1830s. In the background, topped by a cross, is the landmark Silla de Paita. Drawn by Victor Adam, lithograph by Bichebois, in A. de la Salle, *Voyage autour de monde...sur la corvette* La Bonite, *album historique*, Paris, 1851. (Courtesy the George Peabody Library of The Johns Hopkins University)

Plate 8. European and North American observers commented on the sparse furnishing of Peruvian dwellings. This house in Paita contains little more than a hammock for sleeping, a variety of storage and cooking containers, a table for dining, and a chest that also serves as a seat. An open cooking fire burns on the dirt floor. Drawn by Victor Adam, lithograph by Bichebois, in A. de la Salle, *Voyage autour de monde...sur la corvette* La Bonite, *album historique*, Paris, 1851. (Courtesy the George Peabody Library of The Johns Hopkins University)

Plate 9. A Peruvian water carrier. Lithograph, Lima, ca. 1860. (G.W. Blunt White Library)

Plate 10. Like water, firewood frequently had to be delivered by mule to the communities along the arid Peruvian coast. Lithograph, Lima, ca. 1860. (G.W. Blunt White Library)

Plate 11. Europeans and North Americans considered Peruvian women uninhibited and voluptuous, as evidenced by their love of dance. Lithograph, Lima, ca. 1860. (G.W. Blunt White Library)

Plate 12. The 295-ton whaling bark *Clara Bell* of Mattapoisett, Massachusetts, called at Paita three times between 1859 and 1863. Built at Mattapoisett in 1852, she is representative of the more than 350 American whaleships that called at Paita between 1832 and 1865. Paita was visited more than 1,000 times by these American whaleships, with their 30-man crews, during this 33-year period. Sketch by American whaleman Robert Weir. (G.W. Blunt White Library)

Plate 13. Captain Grafton Hillman, a fugitive American whaleman in Paita, advertised his services to the whaling fleet in New Bedford's whaling newspaper, *The Whalemen's Shipping List*, during the mid-1850s. Hillman's role in the dispute over American Consul Charles Winslow is described in chapter 9.

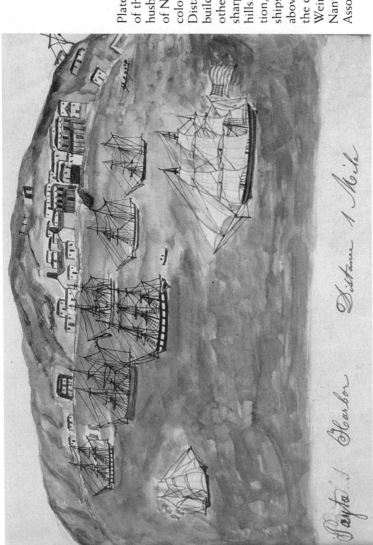

Plate 14. Susan Veeder's journal of the 1848-53 voyage of her husband's whaleship, the *Nauticon* of Nantucket, includes a watercolor titled "Payta Harbor Distance 1 Mile." It shows several buildings painted bright blue and others white, all standing out sharply against the bare brown hills. At right is the signal station, for communicating with ships, and the three crosses above the town probably mark the cemetery on the *tablazo*. (Jack Weinhold photo, courtesy Nantucket Historical Association)

5

The Economic Impact of Whaling on Paita

A lexander Ruden's arrival in Paita around 1836 coincided closely with the beginning of what Alexander Starbuck described as the "boom" in whale fishing in the Pacific. An analysis of the data derived from consular returns from Paita for the period 1832-65 confirms that the decade 1835-46 was indeed a bonanza period for the New England whaling industry. This data also demonstrates the importance of Paita as a port of call for American whalers, and strongly suggests the scope of the economic impact of the United States whaling industry on Paita.

Many of the New England whalers that called at Paita to reprovision or to off-load oil ranged far out into the Pacific, calling also at ports in Australia, New Zealand, and the Pacific island groups, including Hawaii. Most, however, concentrated their efforts closer to the South Ameri-

65

can continent, fishing in the waters of the Galapagos, Easter Island, and Juan Fernandez Island, and thus called more frequently at the Peruvian port.[1] Several ships called at Paita repeatedly during the period 1832-65, suggesting that they found it a convenient port of call, with cooperative authorities, where provisions and marine stores were adequate.

The average length of a whaling cruise, from the time the ship left New Bedford or Nantucket until it returned to home port, was from 24 to 36 months. Obviously they had to reprovision, take on water, and undergo minor repairs several times during that period. In addition, judging from the consular returns, others called at Paita to off-load and cooper oil prior to continuing their cruise. Some of the ships that called at Paita had full cargoes of oil and used the port call to reprovision prior to the long haul to Talcahuano, in southern Chile, the last port of call before rounding Cape Horn on the way home.[2] Unfortunately, the consular records for Paita only hint at the nature of the port calls by American whalers, whether they were merely to provision, whether they were to off-load oil, or whether they were on their way home. This does not, however, diminish the importance of these calls to the local economy, since virtually all the American ships that put into Paita probably purchased food, water, supplies, and marine stores.

The Rise and Fall of Whaling

As might be expected, there are some gaps in the information on the impact of the whaling industry on Paita for the early period (1832-39), when the Englishman Charles Higginson was acting American consul in Paita. Perhaps a regular system for gathering and submitting information to Washington had not been developed, or perhaps some of the consular returns for that period have disappeared. Moreover, for this early period the only data available is

THE ECONOMIC IMPACT OF WHALING

on the number of United States ship calls and the size of sperm oil cargoes. The data on the *value* of oil cargoes only begins in 1839.

The reconstructed data for 1835 suggest that the number of United States ship calls at Paita that year (119) surpassed that of any other year in the one-third of a century for which reliable data are available (see figure 1).[3] Only a half-dozen or so of the American ships that called at Paita that year were not whalers.

At the same time, figure 1 clearly shows that in 1835 the *volume* of sperm oil aboard these ships was the result of an even greater rate of growth. Since the price of sperm oil in New England ports remained fairly constant throughout the mid-1830s (see figure 2), this suggests a substantial increase in United States demand for sperm oil during this period.

Figure 1 also shows that the impact of the 1837 depression in the United States was only felt in Paita two years later (1839). The 1837 panic was caused by the federal government's suspension of specie payments, the proliferation of state banks and note issues, and the subsequent failure of many of those banks. As a result, there was wild speculation in commodity markets, including sperm oil.[4] In 1839, the year in which data for American ship calls at Paita and volume of sperm oil cargoes reached a nadir, whale oil prices in the United States were higher than at any time during the period 1832-49. Data on the impact of the American whaling fleet on Paita in 1840, however, suggests that the industry had experienced a substantial recovery, only to fall again to what appears to be more stable levels throughout the rest of the decade.

The data run contained in consular returns from 1839 through 1854, at which point the consulate changed hands and some of the consular returns seem to have been lost, is more extensive and more reliable. In addition to ship visits, tonnage figures,[5] and sperm oil cargoes, the returns from Paita include the declared dollar value of each ship's

67

oil cargo and - beginning in 1841 - take note of relatively small amounts of right whale oil on board some ships, in addition to the more valuable sperm oil.[6]

The data for the 1840s show interesting correlations between the number of ship calls at Paita, the volume of sperm oil cargoes, and the value of those cargoes. While the number of ship calls remained fairly constant for 1840-42 (67, 59, 67), the *volume* of sperm oil on those ships dipped sharply from a high in 1840 and, after a very modest recovery, in 1842 was on the brink of falling even further. The *value* of that oil followed a similar pattern until 1842, when it too took a dramatic plunge roughly parallel with that of the other two variables. Although United States sperm oil prices experienced a modest decline during this period (see figure 2), part of the the significant fall in the volume and value of sperm oil carried on ships calling at Paita during the 1840s may have been due to changes in whaling patterns. The increasing scarcity of sperm whales along "the line" (equator) in the Pacific, coupled with an increasing industrial demand for "whalebone" (baleen) meant more whalers ventured into the North Pacific in search of right whales during the 1840s and through the Bering Strait into the Arctic Ocean after 1848.

The period from 1843 to 1854 was one of relative stability as far as the number of United States ship calls at Paita was concerned. But the discovery of gold in California in 1848, while not dramatically affecting the *numbers* of United States ships calling at Paita, did radically alter the *ratio* between whalers and merchantmen. This is demonstrated by the dramatic drop in both the volumes of sperm and whale oil, and by the decline in the value of oil cargoes, beginning in 1849. The consular returns for that year show, for the first time, United States ships in ballast calling at Paita. In 1851, 27 ships in ballast touched there, along with ten merchant ships. Most of them were probably bound for San Francisco with a full load of immigrants and merchandise for the gold fields of the Sacramento River valley, or returning from San Francisco and looking for a cargo of guano or some other commodity to carry back to the Atlantic.

The gold fever had a detrimental impact on the whaling industry in general. According to Starbuck, even though shipowners offered generous wages, some men would ship aboard a whaler as a cheap way to get to California, jumping ship when they arrived in San Francisco. Even masters and officers abandoned ship during the highest pitch of the gold rush. These desertions seriously crippled the efficiency of the ships, some of whose masters apparently yielded temporarily to the temptation to turn their whalers into cargo and transport ships.[7] Nevertheless, the industry was probably already overextended, the whale stocks were overexploited, and investors were already redirecting their whaling profits in other directions. The evidence suggests that internal forces were at work beyond this external lure for the labor that was not adequately compensated in the whaling fleet.

Despite the relative stability suggested by the data on ship calls and volume of sperm oil cargoes for the mid-1840s, the curve representing the *value* of oil cargoes rose again rather dramatically from 1843 to 1849, reflecting another brief era of high prices and, by extension, high profits for the Pacific whaling industry. This bonanza began to crumble in 1849, however, and by 1854 the volume of oil and the value of cargoes had reached an all-time low.

After a 39-month hiatus, regular consular returns for Paita once again became available in late 1857. The semi-annual reports of earlier years were replaced by quarterly reports, which contain much more information about the individual American whalers and merchantmen calling at Paita.[8] The National Archives has these more complete reports from the third quarter of 1857 until 1865, after which the returns become very bare-boned, listing only the ship name, tonnage, and volume of sperm oil cargo. Although there are a few gaps in the documentation, it is possible to

69

construct a generally accurate image of the way in which the whaling industry affected Paita through the year 1865, which I have chosen as a cut-off date because of the dramatic decline in the number of ship calls at Paita.

Returning to figure 1, the most dramatic economic indicator derived from the consular returns for this latter period is the spectacular rise in the value of sperm and whale oil cargoes aboard American ships calling at Paita in 1857 and 1858, especially when contrasted with the all-time low registered in 1854.[9]

Ship arrivals dipped slightly in 1859, shot up again the following year, and then in 1860 began a steady decline to nothingness. The *value* of oil cargoes from 1860 on closely follows the same pattern. By the end of 1865 the number of United States ship calls at Paita had dropped to the lowest point in a third of a century (12), even lower than the previous nadir (27) of 1854. And one-third of these were cargo ships carrying coal and lumber from California and Oregon. Similarly, the volume of sperm oil carried on the eight whalers that called at Paita (right whale oil figures cease in 1860) dropped to an insignificant 2,450 barrels, whose value is not even listed on the consular returns.

By the late 1860s, consular returns from Paita do not even list the volume of oil carried by the few American whalers that stopped in Paita, further evidence of the dramatic decline of the whaling industry's impact on the port. In 1868 only 30 American ships, mostly whalers, called at Paita, and during the fiscal year ending 31 September 1871 only 22 American bottoms called at the port.[10]

The effect of the Civil War on the whaling industry was dramatic. According to Starbuck, "no special commercial interest was in a poorer state to withstand war than the whale-fishery." After the outbreak of hostilities in 1861, the New England whalers in the Atlantic were prey to the Confederate commerce raiders. Whalers in the Pacific did not know whether to risk returning to Nantucket and New Bedford, in order to sell their cargoes, or to stay in foreign ports. Many of those that did return were captured and burned. In June 1865, without knowledge that the war was virtually over, the Confederate steamer *Shenandoah* captured and burned two dozen Yankee whalers near the Bering Strait. Starbuck estimates that over 30 ships and barks were captured or burned by the Confederacy during the war. In addition, some 40 idle whalers (the "stone fleet") were purchased by the United States government, filled with ballast and sunk at the entrance to Charleston and Savannah harbors early in the war.[11]

Overfishing - the "scarcity and shyness" of whales in Starbuck's words - and the discovery of a cheaper alternative to whale oil were instrumental factors in the decline of the Pacific sperm whaling industry and its much diminished impact on Paita. The first commercially drilled oil well was sunk near Titusville, Pennsylvania, by Edwin L. Drake in August 1859. News of Drake's success encouraged others to follow suit, initiating a small petroleum boom in the United States at the outbreak of the Civil War. The crude oil was refined into kerosene in existing coal oil refineries, and by the early 1860s kerosene lamps had almost completely replaced sperm-whale oil as a source of domestic illumination.

Of course Paita was not the only port affected by the decline in whaling. Whereas in 1840, 29 American ports had sent to sea 254 whaling vessels, just 17 ports sent a total of 136 whalers to sea in 1866. Already in decline due to its isolated location, Nantucket's outward-bound fleet had diminished from 25 vessels in 1840 to just four in 1866. New Bedford's whaling fleet had reached its peak in the 1850s, but it too felt the dual effects of the war and the widespread adoption of kerosene as a source of illumination. New Bedford departures, which totaled 71 in 1840, numbered 55 in 1866. The once powerful New England whaling fleet, which in 1846 consisted of 735 ships, was reduced by 1875 to 119 ships and barks, and 44 brigs and schooners, with a total tonnage of 37,733.[12]

CHAPTER 5

Quantifying the Economic Impact

I can only speculate as to how the dynamics of the Pacific Ocean whaling industry, as reflected in the tabulated data from consular reports, impacted on the port of Paita. The data suggests that Starbuck's "boom" decade really lasted until 1848, and that it might have continued were it not for the discovery of gold in California. They also suggest that the merchants of Paita, the farmers of the Chira River valley, the water carriers of Colán, the firewood gatherers, and the local industries that produced naval stores must have flourished during this 15-year period, and during the similar but shorter boom that began in the mid-1850s.

In order to accurately describe this economic impact, it would be necessary to assign a monetary value to the water, food, provisions, and services provided to the American whaling fleet during the period 1832-65, which in turn would suggest the volume of hard currency that entered the Peruvian economy through Paita's service sector. A number of obvious and insurmountable problems make this an impossible task. Firstly, taverns, boarding-houses, and bordellos do not usually give receipts. Then there is the problem of how much drinking water was acquired in Paita and what it cost. The third problem that arises is that of trying to estimate the value of ship repairs and other maritime services provided in the port. Finally, and most importantly, the total value of the food and other provisions shipped aboard the 1,500 or so New England whalers that called at Paita would have to enter into the calculation.

With regard to this final point, extant documentation found in the manuscript collections of the New Bedford Whaling Museum in New Bedford, Massachusetts, and the Kendall Whaling Museum in Sharon, Massachusetts, does shed some interesting light on the economic impact of the whalers on the economy of Paita and that of the neighbor-

THE ECONOMIC IMPACT OF WHALING

ing port of Tumbes.[13] Admittedly, the data base is extremely sketchy. I found, in these collections, 34 separate receipts or ship account book entries documenting purchases in Paita during the period 1835-63. The sales receipts, many of which are written in Spanish, are from individual suppliers, local commercial houses, or ship agents. The amounts range from $12 to $537.40.

Similarly, I found 49 individual receipts or account book entries for the port of Tumbes for the period 1841-63, ranging in value from $41.22 to $520.15. Moreover, during this same period I also found six individual receipts or account book entries that group together purchases made in *both* Paita and neighboring Tumbes at approximately the same time, without specifying what was purchased in each port.

The potential data base, then, consists of 89 receipts and accounts from 16 different whaleships, or approximately 6 percent of the estimated 1,500 ship calls in Paita during this period. But considering the time that has elapsed since these accounts were generated, and the fact that they survived the perils of long sea voyages, it is rather remarkable that we have primary sources upon which to base any calculation. In view of the paucity of primary data, I decided for several reasons to include data from purchases in the whaling port of Tumbes in this analysis of the "small trade" generated by whalers. The two ports served the whaling industry in almost identical fashion. They are similar in climate and geography, although Tumbes is more tropical and was far less healthy than Paita. With a handful of exceptions, the provisions and supplies purchased in Tumbes were identical to those acquired in Paita, and in general prices seem comparable.[14]

The data base shows that the average value of the individual purchases made in Paita between 1835 and 1863 was $118.62 - over $2,000 in 1991 dollars. The average value of the purchases made in Tumbes between 1841 and 1863 was considerably higher - $242.46, or over $4,100 in 1991 dollars - primarily because in the latter port drinking water was free for the taking and because it was the market of preference for purchasing firewood and sweet potatoes.

Combining these two averages with the data from the six accounts of purchases made *both* in Paita and Tumbes mentioned above gives an average illustrative value of $190 for the purchases made during each of the 89 documented ship calls between 1835 and 1863. A detailed examination of the consular returns from Paita for the period 1832-65 shows that at least 457 identifiable American whaleships and merchantmen made a total of 1,164 port calls at Paita during this third of a century. If, in addition to this number, we estimate the ship calls for the nine years for which no consular reports are available, the number of probable calls by American vessels increases to at least 1,500, or an average of approximately 45 per year. If we multiply $190 by 45, we arrive at an illustrative value of $8,550 - over $145,000 in 1991 dollars - for the annual economic impact of the "small trade" on Paita for the period 1832-65. While clearly not a fortune in contrast to the money to be made from whaling, that is a lot of potatoes and onions.

Perhaps more important than this exercise in nineteenth-century price history, the whaling papers in the New Bedford and Kendall museums provide interesting data on the kinds of provisions purchased in northern Peruvian ports in the nineteenth century. In terms of both volume and value, the most important items purchased in Paita and Tumbes were "Irish" (white) potatoes, onions, sweet potatoes (*camotes*), beef, and firewood. The onions provided an important source of vitamin C. Mutton, kid, pork, and barnyard fowl also figure prominently on the shipmaster's list. Many accounts include a general item labeled "market bill," representing a variety of seasonal fruits and vegetables purchased in the public market. Other accounts itemize such vegetables as fresh and dried corn, beans, squash, watermelons, oranges, and limes. Rice, salt, sugar, molasses, and pepper figure on many lists, and others include milk and

THE ECONOMIC IMPACT OF WHALING

eggs, obviously for immediate consumption. Finally, many of the accounts also include small purchases of cacao, coffee, wine, brandy, and rum, along with locally produced medicinal supplies such as sarsaparilla root and bark, and tincture of guaiacum. The items described above suggest that, while a sailor's diet might be monotonous and unbalanced at sea, while in port and during the first weeks of a cruise a wide variety of fresh food was available in the crew's mess and at the captain's table.

The same whaling papers also contain periodic references to payments for marine supplies like oakum, lumber, and beeswax (to wax the twine used in mending sails), as well as a variety of port services, including medical attention, blacksmithing, carpentry, caulking, drayage, and the fees paid to "landsharks" for watching and rounding up sailors on liberty, to prevent them from deserting. This data is far too sketchy, however, to allow even an educated guess as to the global value of these kinds of goods and services.

Despite the differences in time, scale, and focus, the multiplier effect of the "small trade" on Paita's economy described above must have been similar to that observed by the Basque mineralogist Fausto Elhuyar (1757-1833), one of the great contemporary experts on mining during the late colonial period, in his *Memoria sobre el influjo de la minería en la agicultura, industria, población y civilización de la Nueva España en sus diferentes épocas.*[15] Paraphrasing Elhuyar's analysis to apply it to Paita, the "prodigious" increase in silver production (maritime traffic) in Mexico (Paita) during the late colonial period (from 1830 to 1865) meant a proportional increase in all the "maneuvers and operations" related to mining (the maritime economy). It meant an increase in population growth, agricultural production, cattle raising, and the exercise of all of the *artes y oficios* (including the oldest of *oficios*), causing the region to begin to emerge from the "stationary state" in which it had previously languished.

CHAPTER 5

Inversely, the declining number of whalers that called at Paita from 1848 to 1854, and especially after 1863, meant less business for the purveyors - foreign and Peruvian - of marine stores, food, water, firewood, and other supplies to the fleet. It also implies a decline in the number of local jobs related to the whaling industry, such as ships carpenters, blacksmiths, riggers, and the stevedores who handled barrels of oil for transshipment.

It meant less business for the boardinghouses, bars, restaurants, and prostitutes that catered to the fleet, and declining income for the consulate based on consular fees. As is evident in chapter 9, the demand for medical attention for American seaman also declined sharply. However, as we shall see in chapter 10, a gradual increase in general exports and imports through Paita may have partially compensated for this decline in the New England whaling industry's economic impact on Paita. But obviously by 1865 the golden age of Paita as a service economy for the American whaling industry, with all of its social and economic implications, was a thing of the past.

6

The Social Impact
of Whaling
on Paita

C learly the impact of the American whaling fleet on
Paita goes beyong the mere numbers of ship calls,
the size and value of their cargoes, and the general
scope of the service economy. An alphabetized index of the
data contained in the consular returns, listing names, tonnage, and dates of call, shows that many of these ships
called at Paita numerous times over an extended period,
almost as if the Peruvian town were their home port. The
record is held by the *George and Henry*, with 14 calls during
the period 1840-58, followed by the *Golconda* with 13 calls
between 1835 and 1862. The *Balaena, Canada, Congaree*, and
Hesper all made at least a dozen calls at Paita during this
third of a century, followed by the *Constitution, Pacific, President*, and *Sappho* with 10 or 11 port calls apiece.

CHAPTER 6

Going beyond the statistical reality that the consular records demonstrate, it seems worthwhile to speculate on what the records do not show. The masters and officers of the ships that called frequently at Paita undoubtedly developed business and personal relations with the port's expatriate community, including the American consuls resident there, and with local officials and Peruvian merchants. The crew members probably also developed a set of acquaintances there. These contacts, over time, may have developed into intimate - albeit intermittent - domestic relationships with Peruvian women that would have resulted in the birth of children.

Measuring the Social Dimension

It is difficult to quantify the social impact of the American presence in Paita during the whaling years. Obviously, the port's bustling economic life attracted many foreigners like Alexander Ruden, who settled there and went into business. Some of them married locally and remained. Others returned to the United States or Europe, and some moved on to greener pastures in Lima or elsewhere in Latin America. Many, as a result of consensual unions or casual sexual encounters with the women of Paita, probably left behind descendents, thus contributing to the diversity of the area's gene pool.

As we shall see in chapter 8, the presence of foreigners influenced the quality of life in Paita in many ways. Port facilities, public buildings, and communications were improved, a wide variety of imported goods was available, and the quality of medical care in the region probably improved as a result of the presence of a consular hospital.

Nevertheless, in the absence of reliable census data, the presence of foreigners in Paita and the Piura hinterland is difficult to document in a systematic way. The first real Peruvian census was taken in 1861. Prior to that, in 1836

THE SOCIAL IMPACT OF WHALING

and 1850, extracts from the *matrícula de contribuyentes* gave a general idea of population size, but did not list foreigners in a separate category.[1] Anecdotal evidence suggests, however, that significant numbers of Americans and other foreigners left their mark on Paita in a variety of ways.

The immigration of Americans and other foreigners was made possible by the liberal decree of 17 October 1821, which offered them free entry, equal protection under the law, and liberty to exercise their profession.[2] Beginning in 1828, however, Peru was seized with a xenophobic protectionism aimed primarily at foreign merchants and traders established in Lima. The Salaverry government restored import bans on coarse North American textiles and flour and raised protective tariffs as high as 300 percent. Other nontariff barriers were also created. Navigation laws closed many ports to foreign ships - whaleships were the only exception - and some Peruvian firms were granted what amounted to commercial subsidies or monopolistic privileges. Rights and guarantees earlier granted to foreigners were repealed, and the government's perennial balance-of-payments crisis prompted emergency revenue measures that had a profoundly protectionist impact.[3] In the wake of these measures - which were of course blunted by the turmoil of the times and the weakness and vacillations of the state - in 1838 local authorities in Piura ordered foreign commercial houses to close.[4]

But Lima's control over the far north was at best tenuous. A French traveler who visited Paita in 1837 estimated that the population was about 4,000, and that there were between 50 and 60 Americans and Europeans resident there. Two of the Europeans were French merchants, no doubt agents of French commercial houses in Lima engaged in the petty trade in luxury goods.[5] Undoubtedly, since most of the expatriate community of Paita was involved in one way or another with trade, it was a very transient population. They operated in a political environment

CHAPTER 6

that was at times hostile, at times liberal, and often unstable and violent. In mid-1839 an American naval officer reported that the only American in Paita was Mr. Ross.[6]

Some of these foreign residents of Paita and its hinterland left behind a record of their presence there. The English consul Charles Higginson, for example, lived in Paita for 40 years. His son, also Charles, had a business in Piura. When the senior Higginson retired he returned to England, but in six months he was back in Paita. On his death he was given the largest funeral that Paita had ever seen.[7] France, Spain, Portugal, and New Granada (Colombia) also maintained consulates in Paita during this period.[8]

As we have seen, the first American consul, Alexander Ruden, began to trade in Paita around 1836. After he relinquished the position of consul in 1853, his firm, Ruden & Company, continued to do business in Paita into the 1870s. In addition to its involvement in the export-import trade, Ruden & Company owned a rice plantation located between the towns of Lambayeque and Chiclayo, which sustained $92,000 in losses when it was sacked by a revolutionary mob in 1868. The firm also acted as Paita agents for the Pacific Steam Navigation Company during this period, but it is unclear how long Alexander Ruden remained directly and personally involved in the firm's activities.[9]

A business associate of Ruden, an Englishman named Gerard Garland, served as acting United States consul for several years in the late 1840s. Garland later acquired a cotton plantation called Monte Abierto situated in a wide alluvial valley on the north bank of the Chira River near Sullana. The property, one of the richest on the river, was part of the old Tangarará estate of the Marqueses de Salinas. Garland was associated with a Peruvian resident in Lima named Pedro Enrique de Arrese y Sanudo, the heir to the *marquesado*, in this venture. Garland married in Peru and remained to establish what was to become a very distinguished family.[10]

THE SOCIAL IMPACT OF WHALING

An engineer and "staunch Union man" from Baltimore named Alfred Duvall was originally contracted by the Peruvian government to undertake some engineering studies in the Department of Piura and served briefly as acting American consul in late 1861.[11] Duvall was later hired by Garland as a technical consultant at the Monte Abierto estate, where he installed pumps and other machinery imported from the United States to lift river water up to the level of the cotton fields. Still later Duvall went into partnership with the British consul, Alexander Blacker, and some Paita merchants in a cotton plantation downstream from Monte Abierto.[12]

Duvall, who was born in Baltimore in 1810, was a member of an old and very distinguished Maryland family. The first Duvall, a French Huguenot refugee, settled in Prince George's County around 1640. He was a surveyor or civil engineer, and was responsible for laying out many of the settlements in Maryland.[13] The Duvalls later gravitated to Annapolis, the state capital, where over the years they intermarried with some of Maryland's best known old families, including the Ogles, Ridouts, and Ringgolds.[14]

Alfred Duvall was trained as a civil engineer. Since a check with the archivist at West Point reveals that he did not attend the United States Military Academy, and since there were no other institutions of higher education in the United States in the 1820s and 1830s where one could formally study civil engineering, Duvall may have trained in an apprenticeship program or he may have studied in Europe. He married in 1842, and 11 years later published a treatise on improving Baltimore's water supply. In 1860, in Baltimore, he published a seven-page pamphlet on the construction of a trans-Isthmanian canal in Panama, probably the result of his personal travels in the region. I do not know whether his family accompanied him to Peru in the early 1860s, or whether he went alone. In any case, after leaving Paita he returned to Maryland, and died in 1882 in Oakland, California.[15]

CHAPTER 6

Another civil engineer named William Stirling, possibly of British origin, was also contracted by the Peruvian government in the early 1850s to survey the Chira River with the idea of building an acqueduct some 50 miles inland to irrigate valley farmland. He later acquired a cotton plantation named Santa Lucia, on the Chira River near the village of Amotape. Santa Lucia, which was about the same size as Garland's property at Monte Abierto, was planted with some 300-400,000 cotton plants in 1863.[16]

The first American doctor to work in the consular hospital in Paita, Dr. Thomas J. Oakford, subsequently served as American consul in Tumbes. He married Alcira Campos of Paita in 1852. When he died in Paita of "debility and nervous prostration" six years later, he left behind three small children.[17] His successor in the hospital, Dr. Fayette M. Ringgold, who later became consul in Paita, also married a woman from Paita, as we shall see in the chapter devoted to him. A subsequent consul in Tumbes, Hiram R. Hawkins, also died in that unhealthy port on 20 November 1866, while the first American consul in Lambayeque, Elisha J. Mix, Jr., of New York, was killed in March 1865 when his own pistol fired accidentally during a struggle with a runaway sailor named Daly.[18]

The British botanist and traveler Richard Spruce (1817-93) also lived in northern Peru during this period. He first went to South America in 1849 with Sir William Jackson Hooker, later Director of the Kew Gardens, and the botanist George Bentham. Spruce discovered many new plants in the Amazon and was responsible for sending cinchona plants to India. He returned to England in 1864 after several years' residence in Piura.[19]

While trade and science brought some foreigners to Paita, it was politics that brought former Ecuadorean president General José María Urbina there in the 1860s. During Urbina's controversial presidency (1851-56) he liberated black slaves, tried to rent the Galapagos Islands to the United States as a whaling station, and consolidated

THE SOCIAL IMPACT OF WHALING

Ecuador's national debt. His agreement with British bond-holders, which would have given them some 2,600,000 *cuadras* of land in the Amazon and in the province of Esmeraldas, provoked the wrath of Peruvian president Ramón Castilla, who blockaded Guayaquil from October 1858 until August 1859, in defense of Peru's claims to the Amazon. During the ultra-conservative Garcia Moreno presidency (1861-75), from his base in Paita, Urbina launched three unsuccessful military expeditions against Ecuador's theocratic ruler, one against El Oro just north of Tumbes, the second aimed at the province of Manabí, and the third at the Canal de Jambelí, near Guayaquil.[20]

Politics also brought to Paita two of the people who were closest to the Liberator of South America, Simón Bolívar. Doña Manuela Sáenz, Bolivar's mistress, lived in Paita for over 20 years after being exiled from her native Ecuador in 1835. There she sold candy, pastries, and tobacco, and died at the age of 58 on 23 November 1856 of diptheria. Shortly before her death, Manuela was visited in Paita by Bolívar's 80-year-old friend and former tutor, the genial Simón Rodríguez, who stayed briefly with her before moving on to the village of Amotape, on the Chira River, where he died in 1854.[21]

The accelerated pace of economic life in Paita during the 1850s and 1860s attracted an even larger number of foreigners. They included: a British merchant named Alexander Blacker, who came to Paita in 1851, was consul around 1863, and later became a cotton planter; an Italian physician named Domingo Davini who practiced medicine in Paita for "several years," and who worked in the mid-1860s in the American Hospital; a rather successful Italian merchant named Francisco Guidini, who was associated with his countryman Davini in a mercantile business, and who purchased the 410-ton steamer *Go Ahead* in Paita in 1864 from a Captain Moses H. Penny; a ship chandler named John Reardon doing business in Paita in the late 1880s; a Colombian English teacher named Marco Antonio

CHAPTER 6

Grisolle who was master of the state grammar school in Paita in the 1860s; a man named John Reid, who was left in charge of the American consulate in March of 1860 during the consul's temporary absence in Piura; a young man from Saxe-Coburg-Gotha named Ernest Samuel Curl, who had lived as a boy in Davenport, Iowa, was a "sound Union man," and who served as a clerk in the American consulate in 1862; one George Laverty, who was manager of the American shipping office in Paita in 1863, and his brother Hiram, who appears as a merchant in 1868; a Nantucket man named W.J. Coffin who was consular clerk in 1863; and other merchants and businessmen with foreign-sounding surnames - Jean Robbins, William Alexander Cox Cole, George Gifford, William Mitchell, William Geddy, P. Hynes, Alfonze Carlin, Manuel Joseph, Ambrose Gonsalve, Rosendo Feijo, Wilber Houston, G.D. Glendenin, Thomas Hall, William Meinen, John Brown, E.H. Fisher, Alexander Bathurst, Henry Dickson, and William H. Perry - who appear in consular records, primarily as purveyors of food, clothing, room and board, and other kinds of relief to destitute seamen.[22]

Some of the Americans resident in Paita during the whaling boom represented the seamier side of United States influence on that port. They included a Martha's Vineyard, Massachusetts, captain named Grafton Hillman, the agent and consignee for many United States vessels that called there, who was a 15-year fugitive from American justice for bigamy and barratry. As we shall see in chapter 9, Hillman was the author of a scurrilous attack on the third American consul in Paita. His principal henchman was the leader of a band of foreign thugs in Paita, a dissolute American sailor named William Giddis, commonly referred to as "One Arm Bill," who spent considerable time in the local jail.[23] One master described him in the following terms: "Their [sic] is one rascal on shore that goes by the name of one arm Bill that entices them to desert by getting them

THE SOCIAL IMPACT OF WHALING

drunk and the story is prevelent [sic] here that Gingold [sic] the consul countenances the same, certainly he does not asist [sic] us to get our men."[24]

Foreigners had settled elsewhere in northern Peru during the years of the whaling boom. The town of Chiclayo in the Department of Lambayeque, with a population of about 8,000 in 1833, boasted an American apothecary with a shop on the main square, and an American carpenter. The druggist had been living in Lima and had married locally.[25] Nearby Lambayeque, the departmental capital, also attracted significant numbers of foreigners during the latter years of the whaling boom. Of these, the man who took over the American consulate there in 1865, after Elisha Mix was killed in a fight with a runaway seaman, was probably the most interesting for the purposes of this study.

Born in Vermont around 1817, James (Santiago) Coke Montjoy grew up in Troy, New York, and studied medicine at "the medical college on Broadway" in New York. Dr. Montjoy was in his early thirties when he first arrived in northern Peru in 1849, perhaps attracted by the prosperity that accompanied the whaling boom or lured to the Pacific by the California gold rush. I do not know if he ever practiced medicine in Lambayeque, but it seems likely. In any case, after he served as interim vice consul for a short period in 1865, the State Department confirmed him as consul in Lambayeque, and he remained in the position for many years. Montjoy was married to a Peruvian woman, operated a rice plantation near Lambayeque, and was part owner - with some German merchants - of a rice processing plant on the outskirts of the city.[26]

The Massachusetts Cotton Pioneer

Of all the nonofficial expatriate Americans living in or near Paita during this period, a Massachusetts man named Frederick H. Dorr is perhaps the most intriguing and, in terms of Peru's economic history, the most significant. Frederick Henry Dorr was born on 2 March 1808 in Roxbury, Suffolk County, Massachusetts, today part of greater Boston. The Dorrs, in the words of one local historian, "occupied a prominent position" in Roxbury. Frederick's father, Ebenezer Dorr, Jr. (1762-1849), was the eldest son of Captain Ebenezer Dorr (1739-1809), a member of the Massachusetts Committee of Correspondence, a participant in the Boston Tea Party, the owner of pew number 36 in the First Church of Roxbury, and a slave owner. Captain Dorr was one of those who suffered most from British retaliation during the Revolutionary War.[27]

Captain Ebenezer Dorr had inherited his father's Boston tanning business and his Roxbury mansion with a four-acre garden, the remains of what had been an extensive estate. A mercantile partnership formed in 1761 with a Boston merchant involved Captain Dorr and four of his seven sons in trade with Europe, China, and South America. The extent of the Dorrs' direct involvement in trade with Peru, however, is unknown. Their ship *Ester* was seized by authorities in Lima in 1823 for alleged violation of trade regulations and fitted out as a war vessel. At the age of 21 Captain Dorr's youngest son Sullivan - Frederick's uncle - became the family's agent in Canton, where he was a pioneer in the fur trade between the Pacific Northwest and China. As an old man, from 1843 to 1846 Sullivan pursued his family's claims against the Peruvian government for losses associated with the seizure of the *Ester*, a period which appears to overlap with his nephew Frederick Dorr's residence in Peru. It is uncertain, however, how the two men's activities in Peru were related.[28]

I have been unable to find evidence of when Frederick Dorr, an unsung American pioneer of the cotton industry, arrived in Peru, why he went there, how long he stayed, or how successful he was.[29] It is highly likely that he first came to Peru on a ship owned by the Dorr family, and he may have been sent there to engage in trade or to represent his family's claims against the Peruvian government.

We know of his activities in the late 1830s because the American consulate was called upon to intervene on his behalf. Dorr operated a cotton plantation and ginning mill called Hacienda Viviate, some 33 kilometers from Paita, on the south bank of the Chira River between Sullana and the village of La Huaca. The hacienda at Viviate, which later became the nucleous of a village, had a population of about 300. Many of the more prosperous merchants of Paita had country houses in and around La Huaca. In May of 1839, while Dorr was in Lima en route to Valparaiso on unspecified business, an officer of the local Peruvian army garrison extorted money from Dorr's administrator, an American mechanic named Roman Gordon, who complained bitterly to the acting American consul.[30]

For reasons that may always remain a mystery, Dorr returned to Roxbury about ten years after the incident at Viviate. He is listed in the *Roxbury Directory* for 1852 as a "laborer" living at May Place. Subsequent directories list Frederick Dorr at the same address, and from 1860 to 1875 he is shown to be working with a Philip Dorr, probably a cousin, at "the rubber factory" in Roxbury.[31]

Frederick Dorr's brief involvement in cotton production in the 1830s in the Paita hinterland is significant because of its timing. Dorr's activity at Hacienda Viviate, and his run-in with local authorities, coincide with the period of xenophobic protectionism that swept over Peru, and with the 1838 order by Piura authorities for the closing of foreign commercial houses. His experiment predates the involvement of Peruvian and foreign entrepreneurs in the cotton industry by at least 20 years. The closest parallels to Dorr's pioneer involvement in large-scale cotton production in the late 1830s seem to be the activity of Gerard Garland, Alfred Duvall, and Alexander Blacker in the 1860s

CHAPTER 6

cited above, and that of a Scottish immigrant named Henry Swayne, who bought the San Jacinto estate in the Nepena Valley north of Lima in 1860.[32] Among the factors that contributed to the economic success of Garland and the others - a fate that apparently eluded Dorr - were a revolution in Peruvian trade policy beginning in the 1840s, and the enormous increase in Peruvian cotton exports to England during the "cotton famine" provoked by the American Civil War.

Lost Demographics

As we have seen, the global economic impact of the American presence in Paita during this period, expressed in terms of Paita's production of goods and services for the whaling industry, is easy to identify. It is also clear that Paita acted as a social bridge between the two nations, a scenario for all kinds of casual encounters and established relationships between North Americans and Peruvians. But an accident of history has made it impossible to quantify the extent of this social impact through contemporary church records.

The parish archives for the church of San Francisco de Paita burned sometime just prior to 1880.[33] Baptismal, marriage, and burial records were irretrievably lost. Extant baptismal and marriage records for San Francisco de Paita date from the year 1880, which is outside of the time frame of this study. Nevertheless, I believe that a careful examination of these lost records would have revealed the kinds of quantitative data needed to sustain the hypothesis about Paita's privotal role as a social bridge between two cultures.

I am convinced, for example, that the lost records would have documented a relatively high incidence of marriages between foreign-born men with Anglo surnames and Peruvian women during the period 1832-65. Tangentially, they would also have revealed a high incidence of legitimate children born to these bicultural unions and baptized in the Roman Catholic faith, since in these matters the mother's religion was and still is determinant.

Somewhat less likely, but still within the realm of the probable, the lost records for Paita might have revealed a high incidence of births resulting from consensual unions or casual encounters between lower-class Peruvian women - those described in extant records as "mistas" - and American seamen. My confidence that baptismal records would reveal the names of foreign-born fathers and the illegitimacy of their offspring is based on the fact that a very high percentage of the baptisms recorded between 1835 and 1890 in Colán, where church records survive, as well as in Paita after 1880, are those of "natural children," that is to say, the offspring of consensual unions in which both partners are listed.

Finally, I believe that lost church burial records for Paita for the period 1830-65 would probably show a notable incidence of foreign-born men and their immediate progeny dying in Paita, since - again - the religion of the wife or mother seems to be determinant at the moment of final reckoning. As we have seen, American seamen dying in Paita, most of whom were probably Protestant, were buried by the consul in the seaman's cemetery and did not fall within the spiritual jurisdiction of the curate of San Francisco de Paita. Of course, this surmisal does not take into account the possibility that many foreign-born men, like Frederick Dorr, returned to the United States before their deaths, while others, like Gerard Garland, moved on to greener pastures.

Regrettably, none of these interesting hypotheses can be proved statistically for want of documentation. Nevertheless, surviving Paita baptismal records for the period beginning in 1880 do show a sprinkling of Anglo surnames such as Higginson, Blacker, Houghton, McCubbin, More, Ward, Taylor, Hopkins, and Wendell.[34]

7

The Second American Consul in Paita

The commercial activities of the first United States consul in Paita were apparently so successful, and in time interfered so much with his official responsibilities, that he was obliged to resign. An American woman who visited Paita in 1850 left the following description of Alexander Ruden's household:

The house is very spacious, constructed wholly after the manner of South American houses. The whole front of the lower part is appropriated to business. A wide and pleasant balcony surrounds the entire house at the second story. Large windows and still larger doors, open upon this balcony, and render it an airy and delightful residence. From this balcony you have a delightful view of the harbor, dotted with ships of almost every nation....Mr. Ruden's household consisted of himself and four gentlemen belonging to the firm. All his servants were male natives and he employed quite a number, with a major domo to superintend them....Mr. Ruden was a bachelor....Eighteen years of his life had been passed in South America where he had amassed quite a fortune.[1]

In early 1848 Alexander Ruden went on an extended leave and left an American named Charles B. Polhemus, who had served as United States Consul in Tacna-Arica from 1837 to 1847, in charge of the consulate. Since his father died in late 1847 or early 1848, Ruden may have returned to New York at this time. In November he again asked the Department of State for leave to attend to "urgent [personal] business," announcing his intention to leave a "partner in my mercantile establishment in Paita" in charge of the consulate. He added that it might be necessary for him to leave Peru for an "indefinite term." Where he went and what he did may never be known, but on 12 January 1851 Ruden finally returned to Paita, after an absence of more than two years.[2]

He did not stay long. By mid-1852 Ruden was apparently again absent, and his business partner in Ruden & Company, an Englishman named Gerard Garland, was preparing and signing the consular despatches in his name.[3] By mid-1853, however, the whaling community was beginning to complain about the consul's prolonged absence from his post on personal business. To many, the logical candidate to replace him was the young physician from Maryland who had recently taken charge of the American Hospital in Paita, Dr. Fayette M. Ringgold.

The Patrician Maryland Physician

Dr. Fayette Monroe Ringgold was an eighth-generation Marylander, descended from Thomas Ringgold who settled in Kent County, on the Eastern shore, in 1650. Prior to assuming his position at the American Hospital in Paita he had been the American consul in the port of Arica, in southern Peru.

THE SECOND AMERICAN CONSUL AT PAITA

Fayette's father, Samuel Ringgold, was a Brigadier General in the Maryland militia, an ardent Jeffersonian and an advocate of universal suffrage.[4] A wealthy and prominent landowner, General Ringgold had settled in Washington County, near Hagerstown in western Maryland, in the mid-1790s. Ringgold's estate at Fountain Rock, six miles south of Hagerstown, at one time consisted of 17,000 acres. Although part of the estate was probably leased to tenant farmers, the federal censuses for 1810 and 1820 show that Ringgold owned 49 and 70 slaves in those years. In Maryland politics, the senior Ringgold was a member of the House of Delegates in Annapolis (1795); a state senator from Washington County (1801-06); and a Justice of the Peace. In addition, he served four terms in the United States House of Representatives during the period 1810 to 1818. Samuel was a vestryman at St. John's Episcopal Church in Hagerstown, where in 1797 he was the pewholder who paid the highest sum for his pew.[5]

Samuel Ringgold, who was born in 1770 in Chestertown, on the Eastern Shore, was married twice. At the age of 22, in Philadelphia, he wed Anne Marie Cadwalader. The daughter of General John and Elizabeth Lloyd Cadwalader, she inherited properties in Philadelphia and in Virginia. In a period of 18 years they had eleven children, six of whom lived to adulthood. Anne Marie Ringgold died in 1811 of puerpural fever, just a year after the birth of her last surviving child, during her husband's first term in the House of Representatives.

On 16 February 1813, the young (43) widower married Marie Antoinette Hay in Washington, D.C.[6] The bride, who was around 23 years old, is erroneously identified in Ringgold family accounts as the granddaughter of President James Monroe.[7] The second Mrs. Ringgold - who was born in 1789 or 1790, hence her anachronistic Christian name - appears to have been from a prominent Virginia family.[8] She was reputedly a woman "of rare beauty and accomplishments, and brilliant in conversation."[9] In 16

93

years of marriage, until Samuel's death at the age of 59, Marie Antoinette gave birth to five children, all of whom lived to maturity and married.[10]

I have not been able to establish officially a birth date or place for Fayette Monroe Ringgold or for his three full sisters, but it appears that Fayette was the last of Samuel's fourteen offspring and that he was born at Fountain Rock.[11] Records from the University of Virginia and the University of Pennsylvania, where he took his M.D. degree, indicate that he was born on 29 March 1825. The Christian name chosen for him - Fayette - suggests that he was born sometime shortly after the Marquis de Lafayette's triumphal return to the United States in 1824, and the widespread "Lafayette mania" that ensued. If the birth date on his 1845 University of Virginia matriculation is correct, Fayette was only four years old at the time of his father's death on 18 October 1829 in Frederick, Maryland.

By the time of his death Samuel had been obliged to sell off most of the farms of his estate. Popular accounts attribute his declining fortunes to his high life-style and penchant for gambling. In his will he refered to "the recent change in the state of my pecuniary concerns," noted that his widow and her children had been "comfortably provided for by real and personal property to a considerable amount," and bequeathed the bulk of his remaining estate, including the property at Fountain Rock, to Mary Elizabeth, an unmarried daughter by his first wife.[12]

Samuel's widow figured in the 1830 Maryland census as a head of household with one male child under five (Fayette), two daughters between the ages of five and 15, and a woman between the ages of 30 and 40, most likely the stepdaughter Mary Elizabeth to whom Samuel referred in his will, who was 32 in 1830.[13] The household also included three male and three female slaves.

In 1832 Marie Antoinette married Robert M. Tidball, a lawyer from Winchester, Virginia, who was eight years her junior. After the sale of Fountain Rock they moved to a house she had built on Washington Street in Hagerstown. Tidball figured as the head of household in the 1840 Maryland Census of Hagerstown, but by this time Fayette was no longer living with his mother and stepfather.[14]

For reasons that are unknown, the Tidball family appears to have disbanded sometime shortly after the 1840 census. A "Mrs. Tidball" was listed in the Washington, D.C., directories for 1843 and 1846, living on 17th Street N.W., between Pennsylvania Avenue and "I" Street, just around the corner from the White House.[15] She probably had the two young boys listed in the 1840 census with her. Their father, Robert M. Tidball, appears in the 1850 census for Hagerstown as a boarder, along with two elderly gentlemen, in the home of Mr. and Mrs. Charles Enochs. One Ringgold family account states that after Tidball died in 1860 Marie Antoinette moved to San Francisco to live with her oldest son George Hay Ringgold, but it seems very likely that she and Tidball separated shortly after the 1840 census.[16]

Indirect evidence suggests that sometime before 1840 Fayette moved to Washington, D.C., where he may have lived with his widowed older sister, Virginia Ringgold Key, who apparently remained in Washington many years after her husband died in 1837.[17] Or Fayette could have lived with his father's younger brother Tench Ringgold (1776-1844).[18] After his mother moved to Washington around 1843, he may even have lived with her for a short time. In any case, Ringgold's 1845 matriculation records for the University of Virginia at Charlottesville list his residence as Washington, and his parent or guardian as "self."[19] In an 1853 report to the Secretary of State he also described himself as a resident of Washington.

I am unable to account for Fayette's whereabouts or activities before he matriculated in Charlottesville in 1845. He remained there for only one year, studying chemistry, medicine, anatomy, and surgery, before transferring to the University of Pennsylvania medical school, where after a two-year course of study leading to a thesis on croup, he was graduated in 1847.[20] If the information on his date of birth is correct, the young physician was 22 at the time of his graduation.

On His Own

Shortly after receiving the M.D., Ringgold shipped as the medical officer of a United States Navy storeship bound for California via Cape Horn. With the approval of his commanding officer, he left the ship at Arica, Peru (now Chile), in early 1849. After practicing medicine there, in September of that year - at age 24 - he was appointed American consul in Arica.[21] This appointment, and his subsequent designation as United States consul in Paita in late 1853, may have been influenced by his family's well-known Democratic connections.[22] Or it may have been due in part to the renown of three older siblings: his half brothers Samuel K. Ringgold (born 1796) and Cadwalader Ringgold (1802) and his full brother George Hay Ringgold (1814), all of whom were prominent United States military officers.[23]

Fayette Ringgold's contemporaries described the young doctor as "a gentleman of education and integrity" and a "kindhearted man." After almost two years in Arica, in May of 1852 he resigned his position "for personal reasons" and announced his intention to move to Paita, where he arrived in early August of 1852 to assume his job as resident physician of the American Hospital in that port. A year later President Pierce nominated Ringgold to replace Ruden as consul in Paita. A $5,000 bond was posted on his behalf on 1 February 1854.[24]

Ringgold's Reign

Shortly after assuming his duties as consul in Paita, Ringgold informed the secretary of state that he did not plan to employ a consular clerk, since the volume of work was "so small."[25] Events seemed to ratify his decision, as the number of United States ship calls at Paita fell from 35 in 1853 to 27 the following year, and the value of the oil on those ships dropped even more precipitously, from $616,905 to about $175,000 (see figure 1). In April 1854 Ringgold reported to the State Department that although navigation and commerce in Paita were not declining, they weren't growing much either. He attributed this stagnation to Peru's almost perpetual "revolutionary state."[26]

Although there are not consular returns for the years 1855 and 1856, the figures for 1857 strongly suggest that, contrary to Ringgold's predictions, the whaling trade in Paita bounced back with a vengeance after 1854. The number of ship calls rose dramatically to 68 in 1857, and the declared value of the oil on those ships hit an all-time high, only to drop again during the next two years (see figure 1). Perhaps it was this increased volume of consular work that, in early 1857, induced Ringgold to appeal to his mother's friend, lame-duck President Franklin Pierce, for an appointment to the vacancy at the American consulate in Valparaiso.[27] Or, more likely, it was financial considerations.

During this period the United States consul in Paita earned $500 a year. By way of contrast, artisans and mechanics working on foreign-operated cotton plantations in the Chira River valley earned, in the 1860s, from $720 to $1,500 a year.[28] Although the Diplomatic and Consular Service Act of 1856 denied to the large majority of United States consuls the right to engage in private employment, under the umbrella of their official position and diplomatic immunity, in Peru this prohibition seems to have been honored in the breach. The same act directed that fees charged for depositing ships papers, issuing certificates, and certifying oaths be submitted to the United States Treasury.[29] Prior to the 1856 act, these fees provided the consul with additional income, but the amount varied widely. During the 12-year period from January 1842 to December 1853 the consul in Paita earned an average of $541 a year from these fees.[30] Although Ringgold's predecessor took full advantage of the provision allowing him to trade, there is no evidence that Ringgold engaged in trade in Paita. As a result, by the time he relinquished the consulate in September 1862, he was flat broke.[31]

One of the factors that may have influenced Ringgold to remain in Paita, despite his strained economic situation, was the fact that he - like his predecessor in the American Hospital there - was married to a Peruvian woman. Mercedes Lañas de Ringgold was the daughter of a Peruvian merchant doing business in Paita, one José Lañas.[32] Lacking baptismal and marriage records for the church of San Francisco de Paita, which were destroyed by fire in the 1880s, I have been unable to determine the exact date of their marriage or the number of their progeny. They had at least one daughter, who traveled to the United States in the 1860s to visit Ringgold's sister Virginia, the widow of John Ross Key.[33]

Complementing the biographical profile of Alexander Ruden in chapter 4, Fayette Ringgold's story points toward several additional tentative conclusions about the background of expatriate Americans living in Peru in the middle decades of the nineteenth century. It suggests that impoverished aristocrats found the security of a government sinecure preferable to involvement in commerce, and used the access to professional training made possible by family connections to prepare themselves for survival in a world far less ordered and secure than the one in which they grew up.

8

Paita's Espacio Económico in the Mid-Nineteenth Century

Beyond the narrow rubric of providing goods and services to the American whaling fleet that called at Paita during the nineteenth century, the *espacio económico* of northern Peru produced and exported large quantities of agricultural commodities. For Paita, as for the rest of Latin America in the mid-nineteenth century, commodity trade was the major factor in its integration into the world economy.[1]

Measuring and accurately depicting this activity during the period 1830-65 is not easy, however, because of its cyclical nature and because of the absence or unreliability of statistical records. The Peruvian government did not begin to publish export statistics regularly until 1891.[2] The absence or unreliability of statistical data was a source of constant frustration to American consuls attempting to comply with instructions from Washington. In 1854 Consul Ringgold responded to a State Department circular requesting economic statistics on trade by noting the "utter absence of all statistical records" in Paita.[3]

While the importance of accurate and complete ranges of statistical data may be obvious, in the case of nineteenth-century Peru no one has yet responded to the challenge of Argentine historian Carlos Assadourian, who wrote: "The quantitative reconstruction of the last century's inter-regional mercantile production and flows must become a priority task if we are to improve our understanding of the contradictions that the process of formation of nation states unleashes and [if we are] to achieve a more sensible and fairer theoretical schematic of the forms and consequences that direct market relationships with advanced capitalistic countries imply for the economic systems of these countries."[4]

Regrettably, as Assadourian points out in the case of Argentina, it is precisely the middle decades of the nineteenth century - between the time when internal *aduanas* with their respective documentation begin to disappear in the 1840s and 1850s as part of the process of centralization and national control, and the creation of national statistics offices by most Latin American governments in the 1880s and 1890s - where the absence of accurate trade and export data is most keenly felt.[5]

Yankee Merchandise

The United States government was also interested in stimulating the market for American goods in Paita and northern Peru, to restore the commercial advantage that the United States had enjoyed in the immediate post-independence years. Contrary to prevailing historiography that depicts - and laments - British commercial domination of early republican Peru, Gootenberg maintains that the United States was Peru's most strategic trade partner in the aftermath of independence.[6]

But throughout most of the period covered by this study (1830-65), the demand for North American goods in the Paita hinterland, beyond the marine stores consumed by the whaling fleet, was rather limited. And most of Paita's merchants believed that there was little chance of increasing it.[7] Ingrained customs and habits contributed to the problem. "Trade is conducted in a very peculiar and slow way," Ringgold commented, "and a large sprinkling of Spanish habits still exist." In an 1854 report he said that direct trade to Paita from the United States was small, with an annual value of only about $100,000. On the average, only one or two direct shipments of assorted American goods arrived in Paita every year, and only one Paita commercial house was involved in the direct trade. The problem, he said, was that imported goods could be obtained from Callao as cheaply or even cheaper than similar goods imported directly from the United States.[8]

A similar report in 1870 echoed and expanded upon these themes. The predominance of European manufactures in the Peruvian market was due, in the consul's opinion, to a variety of factors. The overwhelming presence of the Pacific Steam Navigation Company on the west coast of South America favored the activities of European merchants over those of their Yankee competitors. Since most of Peru's exports went to Europe, and since European goods were better and cheaper than American goods, a circular trade had developed that resulted in the cultivation of European tastes on the part of elite Peruvians. Also, according to the consul, the European merchants in Peru were content with lower profits than were their North American counterparts.[9]

The consular returns for the 1850s and 1860s provide some idea as to the volume and value of goods that arrived in Paita aboard American bottoms, but it must be borne in mind that not all of these cargoes were destined for Paita. In addition to the shipments of "assorted cargo" - some of which were valued in excess of $20,000 - American merchantmen calling at Paita most frequently carried shipments of lumber, flour, bran, and wheat from California. Included in the "assorted cargo" were steam engines and pumps for irrigating the farms along the Chira and Piura rivers and machinery for ginning cotton.[10]

Despite the pessimism of the 1870 report on the market for United States manufactures in northern Peru, and the assumption that the British had a stranglehold on the Peruvian market, there is considerable evidence to suggest that after the "cotton boom" of the 1860s began to fade, the monopoly of European manufactures in the market in northern Peru began to dissolve. From 1865 on, for example, consular despatches from Lambayeque - whose agricultural export economy was very similar to that of the Paita hinterland - list the following kinds of United States manufactures imported into that district:

Representative US Products
Imported into Lambayeque, 1865-72[11]

Cotton cloth
"Yankee Notions"
Shoes for adults and children
Hams, cured meats, and seafood
Butter and lard
Household utensils and furniture
Sewing machines
Kerosene and kerosene lamps
Pianos
Firearms
Agricultural implements and hand tools
Cotton gins and presses
Sugar mills and boilers
Distilleries

PAITA'S ESPACIO ECONÓMICO

Brick-making machinery
Railroad ties and rolling stock
Carts, wagons, and coaches
Lumber
Cement

Of course, the situation in the Lambayeque consular district for the period 1865-72 may not be entirely analogous to that of Paita's *espacio económico* for the preceding decade. The scale of the agricultural export economy in Lambayeque from 1870 on probably exceeded that of Piura for the earlier period, thus generating a greater demand for machinery and other imported products; an active job market for foreign engineers and technicians to install and run the imported machinery; and the foreign currency to pay for them. It was this active agricultural market and the presence of many Americans in Lambayeque that justified the maintenance of an American consulate there from 1860 until 1888. Despite this possible difference in scale and the absence of any monetary value for these imports, however, the Lambayeque data is illustrative, particularly so given the absence of official or consular records on United States imports into Paita during this period.

Exports from the Paita Hinterland

The value and variety of exports from the Piura region far exceeded that of imports. During the period under consideration, the Paita hinterland produced and shipped cotton, straw hats, tobacco, hides, cattle, sugar, chinchona bark, salt, wool, silver, and a variety of exotics for the European and North American markets and for the coastal trade. Most of the Europe-bound goods were consigned to Callao-based foreign mercantile houses.

Given the lack of reliable statistics, it is difficult to quantify Paita's exports during this period in terms of their net value, their relative weight in Peru's total exports, or

103

their destination. At best the available evidence is anecdotal. The British consul in Lima estimated, for example, that total Peruvian exports for 1839 and 1840 were worth about $8.2 and $9.7 million pesos respectively, and that silver specie and bullion accounted for about 70 percent of that total. Gootenberg posits that silver accounted for 80 percent of Peru's exports throughout the postindependence period. The accounts of the British firm Gibbs, Crawley & Company of Callao for the period 1 November 1838 to 31 October 1839 show that the firm exported an unspecified amount of silver bullion and ore, camelid wool, skins, cinchona bark, and straw hats. Of these products, the cinchona and the straw hats are clearly from the Paita hinterland, while the "skins" - if they were tanned goatskins - most likely were too.[12]

But given the predominance of silver exports, it is clear that Paita's primarily agricultural commodities made up a relatively small percentage of Peru's total exports at mid century. Finally, a report prepared by the American consul in 1863 of all the exports of the Paita consular district to New York for the period 1 October 1862 to 30 September 1863 lists the following products:

Exports from Paita, 1862-63[13]

Hides and goatskins	$36,841.66
Peruvian bark	3,007.70
Cotton	7,880.75
Wool	1,568.50
Total	**$49,298.61**

A similar report prepared in late 1871 listed goatskins, hides, cascarilla, orchilla, and cotton as Paita's principal exports, with a total value - for the nine months ending 30 September 1871 - of $213,700 in gold. All of the tanned goatskins went to New York. The drafts that Paita merchants received in exchange for their exports were transferred to accounts in England, where they were used to purchase manufactured goods for the Peru market. All the other exports in 1871 went directly to England, via the isthmian railway in Panama.[14]

After hides and cotton, which will be examined below, the straw hats exported through Paita during this period probably ranked in second place in terms of total value for the period under consideration. The hats were made in Catacaos, an Indian community southwest of Piura, from straw imported from Ecuador. On Sundays, agents from Piura would visit Catacaos to purchase the high-quality hats, which were valued from $15 to $50 each, from the weavers.[15] According to the American consul in Paita, it required "great tact and experience to buy from these Indians, for they are shrewd." He estimated that between $300,000 and $400,000 dollars worth of hats - between $5,100,000 and $6,800,000 in 1991 dollars - were exported annually from Paita, primarily to Chile, California, and the United States. The dynamism and economic impact of the Catacaos hat industry is suggested by the town's demographic growth during this period. According to the American consul at Paita, Catacaos had some 18,000 souls in 1854. Eight years later, the American consul estimated the population of Catacaos at 14,000, but in 1871 his successor put the total at 20,000.[16] In any case, Catacaos was larger than Piura during this period.

Cascarilla, cinchona, or Jesuit bark, from which quinine was extracted, was also an important export from Paita during this period, although the quantity had apparently declined since the late eighteenth century. Haenke reports that, between 1785 and 1788, 60,000 *arrobas* (a measure equalling 25 pounds) of cascarilla were shipped from Paita to Lima. The trees grew in the northern part of the Department of Cajamarca, where it abuts Jaen. While Stevenson believed that the cascarilla exported from Paita was "little inferior to the famous cinchona of Loxa [Loja, the southern most province in Ecuador]," by the middle of the nineteenth century the cascarilla exported from Paita was deemed to be of inferior quality, worth from $12 to $34 the hundred-weight. The merchants in Paita extended credit to the bark traders who gathered cascarilla in the interior. It was then shipped primarily to Lima and Chile, rather than to the United States.

In mid-century, sugar and salt were also shipped from Paita to Chile, in exchange for flour, a residue of the important sugar-for-flour trade with Valparaiso that had its origins in the late-colonial period, and which the elite protectionists of the early republic tried in vain to revive. Of course, the sugar shipped from Paita was only a small fraction of the total volume of Peruvian sugar shipped south, since the small-scale sugar producers in the Chira and Piura river valleys could not effectively compete with the large haciendas lining the coastal valleys from Canete to Lambayeque. In the early 1870s sugar from Paita even found its way to San Francisco, where it was bartered for lumber and flour.[17]

In addition to these better-known agricultural exports, the Paita hinterland produced and exported a number of exotic products for which there was a modest foreign market. They included: cochineal, a red scale insect that was gathered from cactus in the Chira River valley and in the Andean highlands to produce red dye; orchilla (archil), a lichen of the *Rocella* and *Lecanora* genera that grows on algarrobo trees and bushes in the Paita regions, and was exported to Europe in the 1850s and 1860s, where it was used as a deep red or violet dye; concurango (from the Quichua *kunturanku*), the dried bark of a South American vine, which was used as a stomachic; and rhatany (from the Quechua *ratanya*), the dried root of one of two shrubs (*Krameria trianda* and *K. argenta*) which was used as an astringent, and was exported from Paita in the 1830s.[18]

The Paita commercial house of Higginson & Company, directed by the former British consul, and the firm owned by Alexander Ruden held a virtual monopoly on the export of these products, most of which went to Europe. The export of orchilla - "a heavy one" according to an 1863 report by the American consul - was controlled exclusively by Ruden & Company, but by the early 1870s prices in Europe had fallen so dramatically that very little was exported from Peru.[19]

In addition to these exotic commodities, the clarity and medicinal qualities of an oil made from an unspecified "bean" that grew in the Chira valley were praised in an 1854 report by the consul. He noted that American and English merchants in Paita had invested in a company with a Peruvian government monopoly to extract and export the oil.[20]

Cotton Was King

Cotton cultivation has prehistoric origins in the coastal valleys of Peru.[21] Prehispanic textiles woven of cotton and camelid fibers are among the finest examples of the tapestry art known to civilization. During the colonial period coarse cotton thread called *pabilo* was sent from the Paita hinterland to the textile workshops (*obrajes*)of Loja, Cuenca, and Quito, where it was woven into a coarse cloth called *tocuyo*, which was used for clothing and for sails.[22] Although William Bell's thoughtful examination of the cotton export industry began in 1825, when approximately 400,000 pounds of Peruvian cotton were imported into Great Britain, the cotton trade with Europe predated independence.

During Stevenson's 1812 visit to the Casma valley, for example, he noted that a considerable quantity of cotton was produced and ginned in a primitive roller-type gin, after which it was packed by "a powerful screw press" and shipped to Europe. In Lambayeque he also saw an "extensive" mill for ginning cotton, similar to the one at Casma. He noted that large remittances of cotton had been made from Lambayeque directly to Europe.[23] As we have already

107

seen, Frederick H. Dorr operated a cotton plantation and gin at Viviate, in the 1830s. Being a Yankee, it is quite possible that he employed - in addition to the American mechanic mentioned in the consular records - an adaptation of Eli Whitney's new, improved cotton gin, which dates from 1790, and which was in widespread use in the United States and Europe by the 1830s. Although a consular report on the cotton industry written in 1863 suggested that Whitney's invention had not yet revolutionized the Peruvian industry, Spruce's 1864 account of the Chira valley cotton haciendas described steam-driven American cotton ginning machines.[24]

Of the short-staple variety, native cotton plants were perennial in the Paita hinterland. Under ideal growing conditions, they often reached a height of 20-25 feet and produced biennial crops for 10 or 12 successive years.[25] Statistics on cotton production and export in the Piura region are not available or are unreliable for the period 1825-50. It is clear, however, that in the mid-to-late 1830s, as economic conditions began to improve in the Paita hinterland, cotton production and prices rose appreciably. Consul Ringgold, in his 1854 initial report on Paita, said that cotton and tobacco were the only export crops. He also noted the production and export of small quantities of naturally yellow and brown cotton, which was prized for its silkiness.[26]

In the 1850s most of the cotton exported from Paita went to the Mexican ports of Guaymas and San Blas. Spruce estimated that 22,000 quintals of cotton were exported from Paita in 1851. In 1854 Ringgold accused the cotton merchants of Paita of trying to keep production information a secret, but in 1861 Baxley observed large numbers of bales on the wharf in Paita, waiting to be shipped aboard a British steamer.[27]

Large-scale exports of Piura cotton from Paita began in the 1860s, during the European "cotton famine" associated with the American Civil War. At the beginning most of it was cultivated on small plots using traditional methods or gathered from wild plants by the local population and sold to the exporters at six or seven cents a pound. Spruce described the method of production on the small farms around Colán. The women and children harvested the bolls from the cotton "trees," spread the fibers out in the sand, and beat them with switches to separate the seed from the fiber. Very quickly this native cotton was selling for as much as 20 cents a pound.[28]

Seizing the opportunity presented by the cotton famine, in January 1861 Gerard Garland began clearing the algarrobo trees at Monte Abierto, just downstream from Sullana, about 25 miles from Paita. The first cotton seeds were planted in September, and two years later he and his Peruvian partner exported cotton valued at between $10,000 and $15,000 from plants that had been in the ground less than a year. Spruce, who visited Monte Abierto in 1863, described a model agricultural enterprise. The engine house - with its "observatory" on top - the ginning house, the warehouse, and the workshops were built about 100 yards from the Chira, parallel to the river. A deep, board-lined channel provided water to the pumps. Next to the gin was a large cement drying ground for the recently picked cotton. A 25-horsepower steam engine provided power to run the four pumps and the gin. It was fired by algarrobo wood and cotton seed. A lane 50 feet wide, lined with willow trees, led away from the river toward the hacienda house and its gardens. It was intersected by another lane of the same width, with an oval "plaza" at the juncture. The houses and *chacras* of the tenant farmers, who raised garden crops and cotton which they sold to Garland at market prices, were spread out along these two lanes.[29]

At the beginning Garland had no trouble attracting tenant farmers to Monte Abierto. In addition to their cottages and garden plots, peons were paid 5 reales a day. Irrigation workers earned 7 reales a day, while the women and children who picked the cotton at harvest time earned

1 real per *arroba* of picked cotton.[30] But as large-scale cotton cultivation expanded in the Chira valley, labor became scarcer and more expensive. This was due in part to the unwillingness of Indian and mestizo freeholders to work for wages on someone else's land.

At the same time, however, the dramatic decline in the whaling fleet's demand for foodstuffs and services that occurred after 1857 probably forced many peasants who had been at least partially integrated into a cash economy to revert to subsistence agriculture. They also had to curtail their tastes for consumer goods, whose purchase had been made possible by the income earned from the service economy that catered to the whalers. Simultaneously, the cotton boom probably increased land values in the Chira valley, inducing some peasants to sell their plots to entrepreneurs who were interested in putting together large plantations. For these and other reasons the independent small freeholder was lured into the plantation economy, losing his economic independence in the process.

Garland's success at Monte Abierto stimulated Peruvian capitalists, including Don Toribio Seminario and Don Antonio Guerra, to emulate him. They invested in pumps and irrigation projects for their haciendas and experimented with new varieties of cotton seed. The cotton planters found that the Egyptian and Sea Island cottons, when grown in the Paita hinterland, were silkier and produced longer staples than they did in the United States. Moreover, according to one account, the imported varieties tended to become perennial.[31]

Bell asserts that, contrary to a widely held view, Peru's response to the European cotton famine was neither delayed or inadequate. The evidence available from Paita seems to sustain this thesis. Some 3,300 quintals (hundredweights) of cotton were sent to Liverpool in 1862 as an experiment, and in 1864 Paita exported 41,000 quintals of cotton. Prices rose so rapidly in Europe that at the beginning of the "famine" Peruvian cotton was shipped via the Panama railway, even though it was more expensive than the traditional Cape Horn route.

Toward the late 1860s, when production began to recover in the United States, cotton became less of a bonanza crop in the Department of Piura and more of a normal commercial operation. Lower prices in Europe reduced somewhat the profitability of cotton production; yet, despite the heavy shipping costs, Peruvian cotton continued to find a ready market in Europe because of its quality, and because lower production costs - cheaper labor and the absence of pests - made it competitive with cotton grown in the United States. Moreover, Piura-grown cotton remained competitive even after the end of the European cotton famine because Paita's up-to-date port facilities and transportation network, unlike those of other ports, lowered shipping and handling costs. These facilities were a direct legacy of the whaling boom. Further south in Lambayeque, for example, the higher shipping costs resulting from the lack of similar modern port facilities forced most planters to switch from cotton to rice by 1867. Up to 1884 almost all of the cotton shipped from Paita went to England, but in the following year a local merchant experimented with a small shipment to New York.[32]

Although it is not within the scope of this study, it is interesting to note that after the partial collapse of the Peruvian cotton industry associated with the end of the American Civil War and the subsequent War of the Pacific (1879-83), production and export of Peruvian cotton entered a period of modest but sustained growth that peaked during the First World War. The size of the Piura cotton crop did not depend exclusively on demand. In early 1871, for example, after unusually heavy El Niño rains, cotton planters extended their fields to unirrigated land, giving rise to one of the largest crops to date.[33] Total Peruvian production of ginned, clean cotton in 1901 was 10,961 metric tons, but by 1906 that figure had almost doubled. The Paita hinterland was an important player in this market.[34]

With the dramatic decline of the Pacific whaling industry's economic impact on Paita after 1857, the economic organization and focus of the Paita hinterland began to change from a service economy to an export economy. As we have seen, the service economy oriented toward whaling had involved numerous independent suppliers and intermediaries, and thus had a significant economic multiplier effect. By the early 1860s, however, the "small trade" with American whalers - water, food, supplies, and marine stores and services - had all but collapsed. The dramatic expansion in the economy that began at almost the same time, focused primarily on cotton, at first involved many small, independent suppliers, but very soon both production and export came to be dominated by a small group of foreign and Peruvian entrepreneurs. The source of much of their capital and of their prominent position in Paita's commercial community was, in very large measure, derived from their earlier involvement in the whaling industry. And an important source of labor for them was probably the small freeholders who in earlier years had produced garden crops for the Yankee whalers, who ceased being independent farmers and became tenants on the large haciendas and daily wage earners.

9

The Third American Consul in Paita

D
r. Ringgold's tenure as American consul in Paita was interrupted in April of 1859 by the need to return to the United States for unspecified health reasons. By early January of the following year he was back in Paita, but "ill health" obliged him to leave the port again briefly in March. By this time Ringgold's absence from post was beginning to elicit complaints from whalemen and from the State Department.[1]

After news of the firing on Fort Sumter in Charleston Harbor on 12 April 1861 reached Paita, Ringgold traveled to Panama by steamer on "urgent personal business." There, in late November, he requested permission to return to Washington to offer his services to the Secretary of War. Upon returning to Paita in January of 1862, Ringgold learned that Washington had denied his request to return to the United States, and had rebuked him for his unauthorized trip to Panama.[2]

Ringgold's fall from grace was precipitous. According to his successor, who was perhaps somewhat biased, under Ringgold's authority United States interests had languished and desertions from whaling ships were frequent. The consulate had fallen into "disgrace and discredit" because the consul's drafts to pay for the care of sick and destitute seamen had been protested in Washington due to the auditors' concerns over his accounting. Moreover, "the shipping interests [in Paita] had passed into the hands of the vilest hounds that ever infested a community. Masters [of ships] were often insulted by them and dreaded to anchor their ships, or give liberty to their men." Ringgold was blamed for failing to remedy this situation.[3]

Under these circumstances, Ringgold's departure seemed inevitable. He left Paita for the United States on 1 September 1862 aboard the steamer *Peru* for Panama, leaving Mrs. Ringgold to settle their affairs. She refused to hand over the consulate flag, flagstaff, seal, and furniture to her husband's successor, claiming that all of it had been purchased with his funds and was therefore private property.[4]

Dr. Charles Frederick Winslow became the third American consul in Paita upon his arrival at that port on 3 September 1862 aboard one of the ships of the Pacific Steam Navigation Company from Callao, where he had served for several years as the consular physician. Upon his arrival, Winslow informed Washington that the Peruvian government had lost his presidential commission and as a result was unable to issue the execuatur recognizing him as the official United States representative in the Paita consular district. He added, however, that although the prefect of the Department of Piura had refused to allow him to act as United States commercial agent in Paita, the captain of the port had informally acknowledged his authority and was allowing him to deal with American vessels calling there.[5]

THE THIRD AMERICAN CONSUL IN PAITA

Apparently the Yankee doctor was so little impressed with Paita when he arrived in early September of 1862 that he regretted having taken the job. He later confided to the secretary of state that he found the place "most undesirable...one of the most wretched places to live in on the earth, and expensive beyond belief." His only consolation, he said, was in knowing that he was able to help sick and destitute American seamen there.[6]

The Erudite Yankee Doctor

Charles F. Winslow was born in Nantucket, Massachusetts, on 30 June 1811, making him 51 at the time he arrived in Paita. His was a seafaring family. His grandfather Joseph Winslow, a shipbuilder from Taunton, Massachusetts, had been rescued by a British vessel from a French ship where he was being held, and was taken to the island of Bermuda in 1755. There he married a local girl, and there his son Benjamin - Charles's father - was born in 1768, a year before Joseph and his family returned to Massachusetts.[7]

Benjamin, who worked as a blacksmith on Nantucket and acquired considerable property, first married in 1792. On the death of his first wife, Jedidah Hussey, he married Phebe Horrocks (or Horrox) in 1802. Charles Frederick Winslow was the third and last of Benjamin's children by Phebe. He studied medicine at Harvard, graduating with an M.D. in 1834. After two years of postgraduate study in Paris, he practiced medicine at Nantucket and in Boston and became a member of the Massachusetts Medical Society. Winslow married Lydia Coffin Jones of Nantucket in 1839 at the age of 28. She was 22.

Although Benjamin Winslow lost much of his fortune through bad investments in commerce and shipping, his son was able to indulge a passion for travel. As part of a trip around the world, in 1844 the Winslows arrived in Hawaii, where he was employed as the consular physician and surgeon in the whaling port of Lahaina. Their fourth child, Maria Louisa Crowninshield Winslow, was born in 1845 in Lahaina.

From the outset Winslow's tenure at the Lahaina hospital was controversial. The bills he submitted to the consul for the care of sick and injured seamen were unusually high, even taking into account the serious nature of their medical problems and the large number of whaleships that called at the Hawaiian port.[8] By 1851 Winslow had been forced to resign, and he was back in Nantucket, where the last of his five children, Frederick, was born on 1 February.

The discovery of gold in the Sacramento River valley lured Winslow to California, where he penned several interesting letters describing the gold rush, which were published in several contemporary journals.[9] While in California, in 1854-55 he lectured to a temperance organization in San Francisco and to the state legislature in Sacramento on the subject "The Preparation of the Earth for the Intellectual Races," which dealt with geology.[10] Years later he traveled and lived in Europe, where he wrote *Force and Nature. Attraction and Repulsion: the Radical Principles of Energy Discussed in their Physical and Morphological Development*, a treatise on geology published in London and Philadelphia in 1869, which received high praise from eminent European scientists.[11] Winslow's continuing interest in intellectual pursuits is evinced by the letter written to President Lincoln in 1862 requesting permission to absent himself periodically from the consulate at Paita for short periods of time, in order to explore the Andes and the Galapagos Islands.[12]

After returning from California, the Winslows settled in Troy, New York, but apparently his sojourn in Hawaii had whetted his appetite to return, and in 1857 he applied to President James Buchanan for the position as United States commissioner to the court of newly crowned King Kamehameha IV. In support of his petition he cited his prior experience in the islands, his acquaintance with the royal family and the cabinet, and his "very friendly

terms with all the missionaries and the principal merchants." Moreover, Winslow presented letters of recommendation from prominent individuals, including Cornelius K.
Garrison and Elisha A. Allen, the latter minister of finance
to Kamehameha, to further his candidacy.[13] As a consolation, President Buchanan gave him the job in Callao.

Like his father before him, Charles Winslow was a
liberal and charitable man. In politics, he came to sympa-
thize with the newly formed Republican party of Abraham
Lincoln. The consular bond for the position in Paita in 1862
was posted by the governor of Massachusetts, John A. Andrew, an ardent abolitionist elected on the same Republican ticket as Lincoln in 1861.[14] Another prominent person
involved in his appointment was former Congressman
Charles Ingersoll of Pennsylvania, a moderate abolitionist
and defender of the Union.[15]

Winslow apparently shared Governor Andrew's
abolitionist sympathies. In 1863 he sent a despatch to the
secretary of state denouncing the involvement of Americans and other foreigners resident in Peru in the "infamous
traffic" of Polynesian Islanders to work on the coastal plantations.[16] Included in that despatch was a copy of an 1862
letter from a seaman who had been recruited to work on a
vessel that went to capture islanders in the New Hebrides
and New Caledonia. The 1862 letter described the inhumane
conditions of the ships that were sent to contract for indentured labor. In the covering letter Winslow reports that
some 400 "Kanakas" were thus brought to Peru, and that
the sponsors of the expeditions, based in Callao and Paita,
made a profit of $40,000 to $50,000 on their investments
once the immigrants were sold in Peru.[17]

The Decline and Fall of Dr. Winslow

Dr. Winslow's tenure as United States consul in Paita was
as fraught with problems as had been that of his predecessor. For reasons that are not entirely clear from the con-

temporary documents, in early 1863 Winslow's enemies in Paita began a campaign to discredit his management of both the consulate and the American hospital.

The bulk of the charges and slanders leveled against Winslow relate to his stewardship of the consular hospital. As we have seen, the hospital had fallen on hard times as a result of Dr. Ringgold's financial problems with the State Department and with local merchants. By all accounts, however, it was running smoothly again by early 1863. Winslow's problems began in October of the previous year when a Connecticut sailor named G.H. Greene from the whaler *Hector* was admitted to the consular hospital with symptoms of tuberculosis. In a pathetic letter written to his parents in late October, he told them of his condition, described the climate in Paita as good for consumptives, and called Dr. Winslow a "first class physician." A subsequent letter to his brother and sister, written in mid-January of 1863, was much less optimistic. Greene, who wanted desperately to go home, told them that he had had diarrhea for three months and denounced Winslow for charging $1.25 a day for board plus $1.25 a day for medical care for each patient in the hospital, adding "it is in his interest to keep me here as long as he can." In late February Greene's congressman from Connecticut wrote to the secretary of state denouncing Winslow and forwarding the letters Greene had written to his family.[18]

At almost the same time that the Greenes' complaints against the consul were being aired, in early March a broadside printed in Piura entitled "The American Consul in Paita," and signed by "Some South Americans," began to circulate in the port. The pasquinade, which Winslow forwarded to Washington, was written and printed in Spanish, with a very bad gloss in English. It accused the consul of pinching pennies by using a homemade American flag; by keeping his office in a local store - where he sold onions and rancid flour; by moving into the American hospital, and by begging food from ship captains in port, all to the

118

detriment of the dignity of the American presence in Paita. It also accused him of illegally assuming the post of resident physician in the hospital, and of starving the patients there in order to save money to bring his family from the United States to Paita. Finally, it denounced Winslow for using the consular pouch to send private correspondence of American ship captains to Washington and for charging exhorbitant fees for shipping seamen aboard American whalers. At the end of March a ship captain named John H. Morehouse wrote the secretary of state from San Francisco, echoing the complaints against Winslow for charging excessively high consular fees.[19]

Winslow saw the broadside and the denunciations of his stewardship of the American hospital as part of an organized campaign to have him dismissed as consul. He defended his handling of the Greene case and his administration of the American consulate and hospital in Paita in a despatch dated 24 April 1863. He said that he had no choice but to keep Greene - who died on 22 February - in the hospital, in view of his condition, and denied that he had illegally received compensation for working in the hospital. Winslow boasted that "the hospital is pronounced the best on this coast by all who visit it," and cited several Americans who had recently inspected the hospital as witnesses to his administration. The consul went on to inform the State Department that he had recently appointed an Italian doctor "of age and experience" as resident physician, and that he provided free medical advice as needed. The consul attributed the slanderous attacks in the broadside to Captain Grafton Hillman, the renegade American described in chapter 6, who had for several years aspired to be American consul in Paita. Hillman's Peruvian "brother-in-law" was alleged to be the author of the Spanish-language version of the broadside. To support his defense, Winslow forwarded to Washington certified copies of three letters testifying to Hillman's involvement in the drafting of the pasquinade.[20]

CHAPTER 9

Winslow's spirited self-defense against the charges in the broadside and the letters written to the State Department seemed to put the matter to rest, but in September 1863 the whole issue was revived. The State Department received a letter dated 29 September from the retired shipmaster of the *Active*, of New Bedford, accusing Winslow of double-dipping - charging shipmasters three months' wages to support hospitalized sailors and then charging the government for their care and maintenance. "Never," wrote Winslow's denouncer, "did I see so much fraud committed in all the ports on the coast as there have been in this port since that C.F. Winslow came into office."[21]

Winslow responded to this renewed criticism, calling charges leveled against him by Captain Wood and by two sailors named Charles McCollam and Albert Harlow "downright lies." He informed the State Department that the crewmen in question - "drunken sailors and vagabonds" - had feigned illness in order to be released from their contracts and hoped to use the money paid by their master for hospitalization to finance their passage to San Francisco, "the mecca of many a profligate['s] pilgrimage." Moreover, Winslow denounced Captain Wood as a "stupid and superannuated [British] shipmaster out of business and living on charity" in Paita who, despite his anti-Americanism, hoped to become the hospital administrator as soon as his friend Captain Hillman was named consul.[22]

Despite Winslow's blistering self-defense, the State Department apparently felt that enough smoke was coming from Paita to suggest that there might be some fire there. In late October 1863 Washington sent a Dr. James Mackie to examine the situation at Paita. Mackie's visit sometime in November took place while Winslow was absent from post on a "short scientific expedition into the Andes." His clerk, W.J. Coffin, gave Mackie full access to consular and hospital records, but refused to answer questions about Winslow's personal affairs.[23]

THE THIRD AMERICAN CONSUL IN PAITA

Winslow was both offended and worried by the visit of Dr. Mackie, who apparently left no recommendations about the management of the hospital in Paita, and whose written report to the State Department - if there was one - is not in the consular records. Thus, in December Winslow announced his intention to return to Washington to defend himself against all of the allegations about his conduct, making use of the leave of absence granted to him two years earlier. In his stead he appointed a Genoa-born physician with a significant name - Raphael M. Columbus - who had worked in Tumbes for five years as a physician and dentist, and who had on several occasions been acting consul there. Winslow departed Paita by steamer for Panama on New Year's day 1864, leaving Dr. Columbus in charge. In late January he wrote the State Department from Boston, where he was visiting his sick wife, saying that while his critics in Paita "may not have been personally pleased with me," consular affairs there had been conducted with "dignity."[24]

From Boston, Winslow mounted a vigorous effort to vindicate himself. His old friend Governor John Andrew wrote Secretary of State Seward asking him to extend "courtesies" to Winslow. At the same time, Winslow forwarded to Washington an affidavit from the British consul in Paita in which his English colleague responded to Winslow's rather leading questions about the situation in Paita and about his stewardship of the American consulate. The written responses that were most germane to the charges against Winslow tended to vindicate the Yankee doctor. Alexander Blacker, for example, affirmed that American whaleships often discharged sick seamen without observing the formalities of the law, a complaint that is constant throughout the consular records. He also said that since Winslow's arrival in September 1862 there had been a general "housecleaning" of undesirable Americans in Paita, and that the threats to ship captains had diminished. Blacker added that Winslow was "a perfect gentleman," that his character and dignity were superior to those of his predecessor, and that there had been a "very marked improvement" in the consulate's respectability since Winslow took over.[25]

Apparently the case against Winslow was a flimsy one, since sometime in late February of 1864 he returned to Paita to resume his duties. A fortnight later, however, he officially tendered his resignation, citing his wife's delicate health as the justification and expressing his hope that Dr. Columbus would be appointed in his stead. While waiting for official acceptance of his resignation, Winslow traveled at his own expense overland from Paita to Cuenca, in southern Ecuador, and from there to Riobamba and Quito. He intended to observe the volcanoes of the Ecuadorean Andes and "to enlarge the boundaries of knowledge." Dr. Columbus served as acting consul in his absence. Returning to Paita in late June of 1864, in his "trip report" Winslow described an interview with Ecuadorean President Gabriel García Moreno, whose mind was "in such an abnormal state that he cannot be [held] accountable for his atrocities either to man or God."[26]

By mid-July 1864 Winslow still had not heard from Washington about his resignation. It seems that no one was eager to succeed him at Paita. In early May the State Department had offered the job to a Republican office seeker from Syracuse, who declined the honor on the grounds that the annual salary of $500 would not meet his living expenses, and implored Secretary of State Seward to make him a better offer. Impatient at Washington's four-month delay, on 16 July Winslow formally delivered the consulate into the capable hands of Dr. Columbus and embarked on a British steamer for Panama.[27]

A Greenhorn Consul

Although the gentleman from Syracuse was not impressed with the offer of the consulate in Paita in May of 1864, a New Hampshire-born resident of Cuyahoga Falls, Ohio, was. Although he had never been to Peru and probably had very little idea of what to expect in Paita, Henry S. Wetmore readily accepted the Department of State's nomination in September of 1864. After posting his consular bond and receiving his passport, Wetmore took passage to Paita, where he arrived on November first. However, a delay in receiving his exequatur and commission from the Peruvian and American diplomatic authorities in Lima prevented him from formally assuming his post until a month after his arrival. Wetmore only lasted six months as consul. In mid-January of 1865 he submitted his resignation, complaining that the salary was inadequate to cover even half of his living expenses, since he had no other source of income. After sending several increasingly desperate despatches to Washington, on 30 April 1865 Wetmore turned over the consular archives, seals, library, and furniture to one Joseph M. Havens and took the first steamer north.[28]

Havens, who left behind no trace of his presence in Paita, watched the store until early November of 1865, when Dr. Columbus was formally designated as the fifth American consul to serve in the Peruvian port. A quick trip to New York City in late March or early April of that year had allowed Columbus to become a naturalized United States citizen on 8 April, just a week before President Abraham Lincoln was assassinated at Ford's Theater in Washington. On 1 May Columbus embarked on a steamer for Aspinwall, the Caribbean terminus of the Panama railroad, en route to Paita. In August three gentlemen from New York posted his consular bond.[29]

Winslow's Legacy

The evidence available in the consular records for the period of Charles F. Winslow's tenure as United States Consul in Paita suggests that Winslow was primarily interested in using his sinecure as a base of operations for travel and "scientific" explorations, and only secondarily concerned with the operation of the American hospital and the care of sick American seamen. Winslow's frequent self-financed travels, and the fact that his family maintained a separate household in Boston while he was in Paita, suggest that he had an independent source of income, reducing the credibility of the malfeasance charges against him. The final irony of his troubled tenure in Paita, however, was his embarrassment when two drafts for $2,080 and $326 drawn by him on the United States Treasury as reimbursement for the care of sick and destitute seamen were refused by the auditors in Washington, just as Dr. Ringgold's drafts had been protested in 1862. An even further irony was the fact that the holder of these bounced drafts was the Peruvian merchant José Lañas, Dr. Ringgold's father-in-law.[30]

In the final analysis, perhaps the most credible testimony to Winslow's integrity was the revival of the American Hospital under his tutelage and its regular functioning until the unknown date that it closed. Paradoxically, the virtual silence of the consular records on matters of routine hospital administration from early 1863 until the consulate was closed at the end of 1874 suggest that, although reduced in scope because of the decline in whaling, the institution functioned smoothly. Most of the hospital's quarterly reports from this period, detailing the numbers of American sailors treated in the consular hospital, the length of their stay, and the cost of treatment, are missing from the consular records, which make reference to them. The last extant quarterly report, dated 10 September 1863, lists the names of 24 men treated during the third quarter of that year, for a total of 959 man/days at $1.50 per day, for a total cost of $1,438.50. Nevertheless, the consul continued to send to Washington regular statements of accounts current and vouchers for the "relief" of destitute and sick American seamen. Unfortunately, although the annual sums were considerable, these accounts do not differentiate between monies spent to support stranded mariners, and monies spent on medical treatment. A summary for the period 1869-72 shows that the consul spent an average of $3,738 dollars a year for "relief" during this period.[31]

Other indirect evidence of the continued existence of the American Consular Hospital in Paita are the references - in despatches from other posts - to the sending of sick American sailors to the hospital in Paita. The consul in Lambayeque, for example, reported in 1864 that stranded American seamen were "dying of starvation" in his consular district, and that he routinely sent them to the hospital in Paita, even though there was a Peruvian hospital in Lambayeque. Once recovered, the American sailors would have a better chance of shipping aboard an American bottom in Paita than elsewhere on the northern coast. The last such record of the sending of a sick American seaman to Paita from Lambayeque was in March of 1867.[32]

The consular records from Paita do demonstrate Winslow's awareness of Washington's growing preoccupation with the cost of maintaining the hospital in Paita at the time of budgetary problems occasioned by the Civil War. In light of these concerns, and perhaps as a way of deflecting the charges leveled against his stewardship of the hospital, in October 1863 Winslow informed the State Department that he was trying to reduce the cost of maintaining the hospital. Nevertheless, in December he submitted a voucher in the amount of $715.49 for reimbursement of monies spent on clothing for destitute American seamen for the year ending 30 September 1863.[33]

As an economy measure, and in recognition of the dramatic improvements in transportation between Peru and the United States brought about by the introduction of passenger steamers and the completion of the Panama railway, in mid-1863 Winslow requested discretionary authority to send those sick mariners who were able to travel back to the United States via Panama, just as the American consul at Tumbes had been authorized to do earlier. The first seaman to benefit from this new approach was a consumptive with "mild periods of insanity" named John Evans, who was sent from Paita to Tumbes at the consul's expense, and from there embarked on a steamer to Panama. In December of 1863 Washington formally granted Winslow's request. Three months later, however, he informed Washington that he had decided to embark several of the hospital's patients on the United States sloop *St. Mary*, which was due to stop in Paita on its way from Callao to Panama, in order to save the government the $30 that the Pacific Steam Navigation Company charged for a sick seaman's passage to Panama. By the end of 1864, Winslow's successor was able to inform Washington that, in order to further reduce expenses, almost all of the inmates in the American Hospital in Paita had been sent home. He expressed the hope that, once they recovered their health, they could be induced to enlist in the U.S. Navy.[34]

Although the evidence suggests that the Federal government's expenditures for the treatment of sick American sailors declined in the final years of the Paita consulate's existence, Washington continued to express concern at the amounts being spent by the consul in Paita. A report penned by Dr. Columbus in mid-1872 attempts to justify these continuing outlays, and at the same time sheds light on the depression into which the American whaling industry in the South Pacific had fallen. According to Columbus, the consulatate's heavy expenses for "relief" were due in part to the fact that, although very few American ships called at Paita, deserters and sailors discharged from whalers at ports all along the north coast traveled to Paita by steamer or other means in the hope of reenlisting in what once had been the preeminent whaling port of the region. The considerable number of sick or destitute American seamen in Paita was also due to the fact that heavy El Niño rains in early 1871 had increased considerably the incidence of fevers and dysentery among sailors and other foreigners.[35]

THE THIRD AMERICAN CONSUL IN PAITA

In fact, the wave of epidemics that swept down the Pacific coast beginning in 1864 may have counterbalanced the effort to reduce the number of patients at the American Hospital in Paita. In September 1864 the acting consul reported that unusually heavy rains in the north earlier in the year had provoked a "miasma" that had unleashed an epidemic of "bilious fever and dysentery" in Paita. Dr. Columbus said that foreigners were particularly susceptible to the malady - which may have been cholera - and that several seamen had been admitted to the hospital in a "deplorable state." Similarly, in early 1867, Columbus reported that an epidemic of yellow fever thought to have originated in Panama or Guayaquil had resulted in only a few cases of the disease in Paita, due to the strict control of Peruvian sanitary authorities. He added, however, that during the first quarter of the year typhoid and dysentery had a dramatic effect on American sailors. A year later, Dr. Columbus again commented on the prevalence of yellow fever along the Peruvian coast, noting that Paita's dry climate, constant winds, and absolute lack of vegetation had so far spared the port from the ravages of the disease.[36]

In the spirit of his 1862 request to President Lincoln for permission to undertake occasional scientific missions, as his consular duties permitted, Winslow submitted a 21-page report on the "scientific expedition into the Andes" that had kept him away from Paita during the visit of inspector Mackie, just prior to his return to the United States to defend his stewardship of the American Hospital. In his report, Winslow speculated that guano exports from the Lobos Islands, visited in 1852 by Consul Alexander Ruden, could stimulate the economic life of Paita just as the earlier exploitation of the Chincha Islands had stimulated Callao. Winslow also discussed some of Paita's other exports, but most of his report describes expanding cotton production in the Paita hinterland (see chapter 8).

127

CHAPTER 9

Winslow also marveled at the remnants of the prehispanic irrigation systems in the department of Piura, and described a contemporary project undertaken at Carasquilla by a Peruvian named Toribio Seminario and his brother, where six miles of canals were being cut through rock and rocky debris for the purpose of irrigating cotton fields. Winslow described his visit to the farms and haciendas of the "synclinal valley" of the Andes as far as Huancabamba and the headwaters of the Amazon. He extolled the landscape and the fertility of the Huancabamba valley and described the abundant production of cochineal, cinchona, and flax. He noted, however, that transportation costs impeded the full realization of the region's potential and opined that the large amounts of time and money needed for the construction of a railroad over the Andes would probably delay the project for many years. Winslow closed his report with export statistics for 1862-63, which also are examined in chapter 8, and included a sampler of 11 different varieties of Peruvian cotton.[37]

The findings of Winslow's report on his trip to Ecuador to study volcanos, while interesting from a political perspective, appear to be only marginally useful in "enlarging the boundaries of physical knowledge" and are not really germane to this study. Clearly, by sending the State Department detailed accounts of his travels Winslow felt that he was justifying his absence from his post at a time when criticism against him was mounting. The doctor's expeditions in South America, however, failed to satisfy his wanderlust. After returning to Boston in mid-1864, Winslow and his family lived in Europe for four years, where he wrote and published his treatise, *Force and Nature*, based in part on his South American explorations. Then, in the early 1870s Winslow moved to Utah, where he was involved in mining enterprises. His wife died in Boston in 1874, and he died in Salt Lake City in June of 1877 at the age of 66.[38]

10

Progress Comes to Paita

P aita may have seemed like a miserable hole to Dr. Winslow when he arrived in 1862, but in fact the town had gradually improved over the years, due both to the economic and social impact of the whaling boom and to its role in the gradual increase in agricultural exports. This progress was also in part due to the fact that Paita was relatively isolated from the strife, militarism, and revolutionary destruction that shook Peru from the independence era until the early 1840s, when General Ramón Castilla stepped onto the political stage. Between 1820 and 1842, the country witnessed over 14 years of almost constant warfare, interspersed by ephemeral, often violent, military governments.

Paita and the Pax Castillana

Paita's remoteness from Lima helped to insulate it from the chaos of the early republican years. The wars of liberation had little impact on Paita beyond Admiral Cochrane's at-

CHAPTER 10

tack on the port in 1819. During the period of the Peru-
Bolivian Confederation (1837-38) and the subsequent Chilean invasion, Paita remained a distant witness to the con-
vulsions that shook most of the rest of Peru. The 1856 revolt of General Vivanco against the Castilla government,
which like so many of Peru's nineteenth-century "revolu-
tions" began in Arequipa, affected Paita in only a peripheral way. Siding with Vivanco, the Peruvian navy frigate
Apurimac and two small steamers seized Trujillo, San Pedro,
Lambayeque, and Piura, but the revolt was quickly sup-
pressed and the impact of the navy operation on business
as usual at Paita was minimal.[1] Not even the brief (1859-
60) conflict with Ecuador, during which the Peruvian fleet
ineffectively laid siege to the port of Guayaquil, had much
of an impact on nearby Paita.

In a positive sense, however, the politics of this era
did affect the northern port, especially after General Ramón
Castilla assumed power in 1845. Castilla, who ruled from
1845 to 1851 and again from 1855 to 1862, "put Peru back
on its feet," in the words of Jorge Basadre.[2] And, thanks in
part to the whaling boom, Paita was in a good position to
take full advantage of this *pax castillana*. The guano bonanza
that began in the 1840s caused government revenues to
burgeon during Castilla's first term. The budget was bal-
anced, the national debt was consolidated and reduced, par-
liament was reinstituted, and Peru's first railroad from
Callao to Lima was constructed. In 1851 Castilla yielded
the presidency to his elected successor, General José Rufino
Echenique, but unfortunately the three years of Echenique's
presidency were tarnished by blatant corruption and ris-
ing popular dissent.

In 1854 Castilla overthrew his chosen successor in
a coup, and the following year a grateful congress elected
him to a second term. With guano income still swelling the
national coffers - and the pockets of the ruling class - during his second term Castilla pursued an ambitious program
of public works and infrastructure development. During

this period Peru abolished the Indian tribute once and for all and freed black slaves. But in order to resolve the nation's chronic shortage of labor, beginning in 1849 the government authorized the entry of thousands of indentured Chinese workers for the sugar and cotton plantations of coastal Peru.

The nation's exports continued to grow at a modest but encouraging pace, more railroads were authorized and built, communications improved, education at all levels was stimulated, and the apparatus of the state was strengthened and modernized. Unfortunately, however, the guano boom produced no breakthrough in economic growth for Peru. According to Mathew, even though the export commodity belonged entirely to the state, and even though sales and profits were controlled by a state-authorized monopolistic concession to foreign commercial interests, "the treasury remained hard up, agriculture and industry enjoyed no general and sustained prosperity, and in 1876 the government was forced to default on its bonded debt" of some 41.7 million pounds sterling.[3] Nevertheless, when Castilla retired from office in 1862 he left behind a very different Peru than the one that he had inherited 20 years earlier. By the same token, Paita in 1862 was a rather different place than it had been at the beginning of the whaling boom in the early 1830s.

Urban Amenities

Estimates for the population of Paita around the middle of the century vary widely. In 1852 the American consul reported that according to an "official census" the population, consisting of Indians, whites, cholos, zambos, and blacks, was 3,200.[4] In 1861 the American vice consul told a visiting American doctor that the population of Paita was 2,500. But the doctor, who described the town as "not prepossessing," opined that it was probably closer to 5,000. The streets, he noted, were still narrow, irregular, and unpaved. Many of the houses were still built in the traditional *bajareque* style, without windows or chimneys, but there were also rough, timber-framed buildings filled in with adobe blocks, and some houses were made entirely of adobe. The housing of the "elite" and the facade of the church were plastered with lime or whitewashed with a solution made from ground seashells. An American woman with a keen eye for detail, Mrs. Madeleine Vinton Dahlgren, noted in 1868 that the houses in Paita were "admirably suited for the conditions of life [t]here." Squier commented that some of the houses, especially the residences of expatriate agents of foreign commercial houses, were "set out with furniture which would not disgrace a Fifth Avenue parlor," while Mrs. Dahlgren visited a "handsomely fitted-up" clubhouse frequented by foreign merchants. The high-pitched roofs of all the houses were still built with bamboo poles from Guayaquil and thatched with palm leaves.[5]

A subsequent American visitor to Paita, who also estimated the population at five or six thousand, described the same kind of housing, which he characterized as not bad dwelling places. He found the town's marketplace and its school bustling with activity, and said that the streets were as clean as one can expect of a place that is swept constantly by sandstorms.[6]

Despite the material improvements, life for most of Paita's inhabitants did not change radically. The economy was still centered on shipping and the sea. Cockfighting remained the favorite sport of clergy and laity alike, and gamecocks were the "best cared for denizens of Paita." Ordinary folks still spent a lot of time "swinging indolently in hammacks" or celebrating a religious festival with "dancing and low bufoonery in the streets in fantastic dress."[7] In the late nineteenth century ladies from Piura, Ecuadorians from Guayaquil, and Peruvian planters from the Tumbes valley summered in Paita for sea bathing and to escape the "pernicious fevers" of the Gulf of Guayaquil.[8]

PROGRESS COMES TO PAITA

Throughout the second half of the century, Paita's main disadvantage continued to be the lack of potable water. The 1854 proposal by an American merchant to bring water from the Chira valley came to naught, as did President Castilla's determination, in the 1861-62 budget proposal, to bring water to the port. Another project to build an aquaduct in the 1870s also failed. In 1889 the Peruvian government granted a public works concession to an American named Edward Fowkes, a resident of Piura, to bring water 30 miles across the desert from the Chira River, over a relatively level course. But the failure of the House of Barings on the London stock exchange paralyzed the venture.[9]

To encourage trade, the Peruvian government built a wooden pier in Paita sometime in the early 1850s. A decade later, President Castilla's budget proposal for 1861-62 included, as a top priority, the building of a new pier in Paita, along with a floating dock in Callao and a pier in Pisco. An American visitor in 1861 described the existing pier as "substantial." Like Castilla's proposal for an aquaduct, the new pier may have never been built, since in the 1890s the government again announced its intention to build an iron pier at Paita.[10]

A further incentive to commerce, and a marked contrast to the simple *bajareque* and adobe dwellings of the port, was the "commodious" prefabricated English iron customhouse built in Paita in the late 1850s. A two-story structure measuring 60 feet square, it was surrounded by a "neat iron balcony" and surmounted by a cupola or "belvedere" and a flagpole. Behind the customhouse was another prefabricated iron building that served as a "public store."[11]

To improve communications and the transmittal of information, a printing press was installed in Paita in the 1840s. A half century later, an American resident in Piura named Emile Clark installed 800 miles of telephone lines connecting the Departmental capital with Paita and other population centers. Moreover, by the end of the century street railways had been built in both Paita and Piura.[12]

A Trans-Andean Link

Of potentially far greater importance than these modern amenities, however, were the several proposals to build a land link from Paita to Piura, and then over the Andes to the Amazon. This link was seen primarily as a means of easing the transport of cotton and other products from the Chira valley to Paita. In 1863 it cost between 1.1 and 1.4 cents per pound to transport cotton by mule back from the plantations along the Chira to Paita, which was about 10 percent of the cotton's value in the port. Another, longer-range objective was to connect the departmental capital Piura and the lower Piura river valley to Paita. Exports from Catacaos and the lower reaches of the Piura took four days to reach Paita on mule back. In addition, the rail link over the Andean *cordillera* was seen as an alternative to the long sea voyage around Cape Horn or the crossing of the unhealthy isthmus of Panama. Finally, it was envisioned by some as a vehicle for opening up the Amazon basin to world commerce via the Pacific.

The first practical proposal to build a railroad to replace the trail from Paita to Sullana, on the Chira River, and then to Piura dates from the early 1850s, at the time that the rail link between Callao and Lima was being built.[13] The credit for the original conception of a trans-Andean railroad linking Paita to the Amazon, however, may belong to a Peruvian. Around 1843, shortly after the construction of the first commercial railways in the United Kingdom and the United States, a certain Rudecindo Garrido conceived the idea of building a rail link between Paita and San Borja, on the Maranon River. The route that he proposed was virtually the same one favored later by several foreign observers.[14]

PROGRESS COMES TO PAITA

Apparently nothing came of Garrido's proposal. At about the time of the construction of the Callao-Lima railroad, in 1854 the United States consul reported to Washington that an American engineer (possibly Alfred Duvall) had recently surveyed a direct and "commodious" route through the Andes from Huancabamba to the Amazon, and was convinced that a wagon road could be built at comparatively low cost from Piura to the falls at Pongo de Manseriche, on the Maranon.[15] The proposed route would traverse the lowest Andean pass from Panama to Patagonia, which a British geographer described in 1906 in the following terms:

Leaving the Port (Payta), and traversing eastwardly the flat coast-zone [to Piura], the line will reach the Andes, and ascending the western slope will cross the summit [at Huarmaca, 100 miles from the coast] at an altitude above sea-level of about 6,600 feet, by means of a pass which seems almost to have been made by Nature, in order that man might create a way of travel between the world's greatest ocean and vastest river, crossing one of the highest mountain ranges of the globe; for in all the 1,500 miles of the Peruvian Cordillera there is no pass at a less altitude than 13,500 feet.[16]

The accessibility of the Huarmaca pass seemed all the more exciting in view of the fact that at the time this route was being considered the Panama Canal was not yet a reality, and the two existing trans-Andean Peruvian railways (Lima-La Oroya and Mollendo-Juliaca) crossed the cordillera at altitudes of 15,642 and 14,666 feet, respectively. Moreover, according to one traveler's account, a short tunnel through the crest of the *divortia aquarum* at Huarmaca, similar to the Galera tunnel on the Lima-La Oroya line, would reduce the high point on the line to under 5,000 feet.[17]

The contract for the 97-kilometer standard-gauge (1.45 meter) Paita-Piura railroad was let in January 1872. Although it called for completion of the line in 22 months, construction was begun in 1875, and was not completed until 1884, after the 1883 Treaty of Ancón which formalized the end of the War of the Pacific. Instead of going due east from Paita to Piura, the first stage of construction linked the port with the Chira River valley and the town of Sullana. The line eventually was extended to the Departmental capital and the rich cotton plantations of the Piura valley, but the dream of linking the port with the Amazon along the route proposed by Garrido and surveyed in 1854 was never realized. Moreover, in 1890 the Peruvian government signed a contract with its foreign bondholders, renegotiating its debt of 51.5 million pounds sterling. By virtue of this agreement, the infamous "Grace Contract," Peru ceded all existing rail lines, including the one from Paita to Piura, to its British creditors for 66 years. Under British management, in the late nineteenth century the Paita-Piura line was described as "one of the best equipped on the Pacific coast." Also during this period an unconnected six-mile line was built privately to link Piura with the hat-making center of Catacaos, which had a population of 25,000 inhabitants at the time, and whose annual production was estimated to be worth $800,000. Also, around the turn of the century, a 30-mile short line was built to connect the sulphur beds of the Sechura desert to the port of Bayovar.[18]

Paita as Window on the World

Although Paita was imperfectly linked with its hinterland, throughout the second half of the nineteenth century its maritime ties with the world at large were modern and efficient. The packet boats of the Pacific Steam Navigation Company (PSNC), which plied the route from Panama to Valparaiso, began to call at Paita in 1840. The PSNC received subsidies from the British government for carrying official mail and at the same time had a monopoly from the Peruvian government. The frequency of the steamers' calls at Paita varied over time. In 1854 the PSNC had six British-built ships that called on a weekly basis at the major ports of Guayaquil and Callao, and less frequently in the secondary ports like Paita and Pisco.

By 1871, the PSNC served 26 Pacific ports between Panama and Valparaiso, 13 of which were Peruvian. North- and south-bound steamers called at Paita three times a month. The British company's tentacles also extended around Cape Horn to the east coast of South America and Liverpool, as well as via Aspinwall (Colón) in Panama to New York, Southampton, Liverpool, Cherbourg, Le Havre, Bremen, and Hamburg. Another line connected the Pacific ports of South America to San Francisco, via Panama. In 1871 the PSNC's fleet consisted of 33 steamers with an aggregate tonnage of 60,145 tons. The flagship of the fleet - which included the 1,800-ton *Payta* - was the 3,500-ton *Aconcaqua*.[19]

Despite the phenomenal growth of the PSNC over time, one thing that remained constant were the complaints about the service and the high rates charged for passengers and mail. Beginning in the mid-1840s, the American consuls in Paita began to send their despatches by steamer to Panama, for forwarding across the isthmus to the United States, instead of the far more lengthy and tedious route around Cape Horn. Although this route saved a lot of time, in 1845 the consul denounced the "very high" rates charged for mail. His successor concurred, calling the charges for both passengers and freight an "enormous tax on the traveling community." Moreover, a French traveler who arrived in Paita aboard a PSNC steamer called them "floating prisons," noting that even though their construction and seaworthiness might be superior, the interior decor, the food, and the service were terrible, while the officers were "stiff and silent." Later in the century, after the English monopoly over passenger service along the Pacific coast was broken and steamers of other flags offered comparable service, the PSNC's rates were still deemed to be outrageously high. In 1869, for example, the firm announced a *"gran rebaja de*

137

tarifas" (great discount in fares) under which a first-class passage from Callao to Paita sold for 20 *soles*. The reverse trip, however, cost 36 *soles*. In 1871 the last American consul in Paita, echoing sentiments of several of his predecessors, urged Washington to consider subsidizing an American-flag line in service to the Pacific coast of South America, calling the PSNC's rates exhorbitant "beyond all reason."[20]

The Boom Sustained

The alleged high cost of shipping from Paita did not preclude a significant increase in the value of merchandise exported from the port, for both the coastal trade and for international markets, toward the end of the century. Customs collections at the Paita customhouse grew from $37,051 in 1866 to $455,700 in 1890, making it Peru's second most important port after Callao in terms of customs revenues. These figures do not include, however, the duties paid on maritime exports from the port of Tumbes and the Bay of Sechura, or on overland exports to Ecuador. Another source gives the total value of exports and imports through Paita in the early twentieth century as between $1,500,000 and $1,600,000 annually.[21]

Expressed in 1991 dollars, the value of exports for a typical year in the late 1890s is about $44 million.[22] It is clear from the above that, although the quantities and values of the products exported from the Paita hinterland to the coasting trade or overseas may have increased by the end of the century, virtually all the products exported were the very same commodities produced and sold during the whaling boom. The notable exception is, of course, the appearance of hydrocarbons on the list of Peruvian exports.

PROGRESS COMES TO PAITA

Petroleum Pioneers

Northern Peru entered the age of petroleum only a few years after Drake's discovery at Titusville, Pennsylvania, in 1859. In one of history's many ironies, by the late 1860s considerable quantities of petroleum and kerosene for illumination were being imported into Paita and other Peruvian ports, which only a few years earlier had been swimming in sperm oil. But within a matter of a few years, Paita was transformed from an importer of hydrocarbons to an exporter. The first shipment of petroleum from the oil fields of the Talara peninsula was shipped to England late in 1870. A year later, on 29 September 1871, the United States bark *Veteran* sailed from Paita for Liverpool with a load of Peruvian petroleum produced by the Peruvian Refining Company at Tumbes. The firm of James Bishop and Company of New York was one of the principal shareholders in the Tumbes refinery. The only drawback to the petroleum being refined at Tumbes was its high benzine content, which tended to make it explosive. At about the same time, an American who for many years had lived in Lima, Mr. James Thorne, imported, via Paita, American machinery to drill exploratory wells on the Talara peninsula, at Negritos. According to the American consul, Thorne's experiment was a success.[23]

Around the turn of the century, foreign entrepreneurs began to focus more attention on the hydrocarbon reserves of the Talara peninsula, the source of the tar and pitch used by the New England whalers. They included the South American Petroleum Syndicate formed in 1901 to drill for oil at Lobitos, William Deswick of the London and Pacific Petroleum Company who was working just to the south, and Faustino Piaggio, who exploited the Zorritos field to the north. Total production of the Talara fields in 1900 was about 40,000 tons, and much of it was exported through Paita.[24]

CHAPTER 10

Thanks in large measure to the impetus provided by increased economic activity in the Talara oil fields, at the end of the century Paita boasted two hotels, two private clubs, a school, a traditional Peruvian-style church and another one with a wooden steeple like a New England Congregational meetinghouse, perhaps part of the legacy of the Yankee whalers who called at Paita until the 1860s.[25]

The city of Piura, which became the principal cotton mart of northern Peru after the boom years associated with the United States Civil War, also underwent a dramatic transformation beginning in the middle of the nineteenth century, the indirect result of the whaling boom. Its population in 1847, estimated at 10,000, was about 10 percent "white," less than 10 percent black, and the rest Indian or *mestizo*. Progress in Piura's cultural and educational life accompanied demographic and economic growth. The income and property of the Mercedarian convent had been applied to the creation of a public secondary school, the Colegio del Carmen, which began functioning in 1845. Two years later it boasted 120 students of Latin, Spanish grammar, and natural philosophy. A British visitor taking an evening stroll through the streets of Piura in 1847 was struck by the music emanating from numerous pianos. Earlier in the decade a weekly newspaper, *La Chispa*, had been launched in Piura and by 1847 it faced competition from another weekly, *El Vijia*, which carried political news of Peru, South America, and the rest of the world.[26]

Toward the end of the century Piura, with a population of 12,000, had eight churches, the Bethlemite Hospital, two hotels, two clubs, and two theaters. Representatives of German and English mercantile houses were established there, and American commercial interests were represented by the firm of Clark and Houston. There were several cotton ginning facilities in the town and the surrounding area.[27]

11

The Last American Consul in Paita

D r. Rafael M. Columbus, the Genoese-born physician who claimed direct descent from his famous namesake, served the administrations of Presidents Andrew Johnson and Ulysses S. Grant as United States consul in Paita for more than five years. He faithfully submitted quarterly reports on the occasional arrival of American ships in Paita and the "relief" provided to American seamen. He responded as best he could to Washington's growing hunger for trade and economic information about northern Peru. There was never a hint of irregularity or scandal connected with his administration of the consulate.

As a medical doctor, Columbus showed particular concern for the sick American sailors who ended up in sick bay in Paita. The wife of the commander of the United States South Pacific squadron, Admiral Dahlgren, who visited Paita for several days in 1858, wrote in her memoirs that

CHAPTER 11

Dr. Columbus was "much respected" in Paita, and was "reputed to be a skillful physician," particularly in protecting his patients from yellow fever. Mrs. Dahlgren added that Columbus was very attached to the United States, and considered himself an American.[1]

But despite his loyalty and his years of satisfactory service, Columbus's tenure as consul in Paita eventually fell victim to two phenomena peculiar to post-Civil War North American politics: the demand of Union Army veterans for official compensation for their sacrifices, and the wholesale distribution of political appointments to government positions that characterized the Grant administration. Not even the remote, now unimportant consulate at Paita was to escape this rash of job fever.

The Veteran from Brooklyn

Paita's low ranking on the list of consular plums available during the presidency of Ulysses S. Grant is best demonstrated by the fact that the new administration in Washington did not get around to assigning a new consul to the Peruvian port until late 1870, almost two years after the hero of Appomattox Court House was inaugurated. The man favored with the appointment was a Civil War veteran from Brooklyn named John Murphy, who was born in Ireland and immigrated to the United States with his family as a young boy. Murphy joined the Union Army in June of 1862 and served for three years in the 139th Regiment of New York Volunteers. His right arm was shattered at the Battle of Cold Harbor, Virginia, on 3 June 1864. Doctors removed Murphy's right elbow joint and fused his arm bones together. The consular documents in the National Archives written by Murphy himself clearly demonstrate the extent of his physical impairment.

After the war Murphy returned to New York and studied law. He practiced as an attorney and notary, working for New York merchants with commercial interests in

THE LAST AMERICAN CONSUL IN PAITA

South America. At some time in 1870, probably as a result of his war experience, he contracted tuberculosis. On the advice of his doctor, he sought a consular position in a warm, dry climate in order to recover from his pulmonary affliction. The best that Murphy's former commander, General - now President - Grant could offer was Paita, Peru, but at least the climate seemed to be appropriate.

The Senate confirmed John Murphy as consul in Paita on 23 January 1871, but the consular commission sent to him later in the month by the State Department was for Pará, in Brazil - a totally inappropriate climate - and he returned it to Washington. In early March, the problem of his commission resolved, he received his passport and sailed from New York for Aspinwall. Murphy arrived in Paita in late March and on 1 April 1871, formally took possession of the consulate from Dr. Columbus. Before the end of the month he received his exequatur from the Peruvian foreign ministry in Lima.[2]

It was John Murphy's lot to close the United States consulate in Paita, after 41 years of service to the New England whalers and other North American interests on the coast of northern Peru. In 1874, in the aftershock of the panic of 1873, Congress failed to appropriate monies to staff a number of United States consulates. At mid-year the decision to close the consulate at Paita was made and communicated to Murphy, who tied up the loose ends and departed Peru for New York in December, aboard one of the scandalously expensive ships of the Pacific Steam Navigation Company.[3]

Murder on the *Tablazo*

The United States consulate in Paita in the 1870s was probably a very good place to send a wounded Civil War veteran who was seeking a cure for tuberculosis. After the departure of the American whalers there was not much to do, and with a competent consular assistant one could man-

143

age affairs with a minimum of effort. John Murphy's pre-decessor, Dr. Columbus, correctly portrayed the situation in Paita in a despatch to Washington three years before Murphy's arrival: "Formerly Paita was a flourishing little port and its prosperity depended in a great measure on the visits from the numerous American whalers that cruised the Pacific; but since the Civil War in the United States, the whalers have greatly diminished and the prosperity of this place has declined in proportion as the whalers have fallen off. At present trade is very dull."[4]

Once in a while, however, an American citizen got himself into trouble in Murphy's consular district, requir-ing a flurry of activity that would break the tedium of the consul's existence. Such was the case when an American engineer named A.W.D. Forbes, who had been contracted to work on Henry Meiggs's Lima-La Oroya railroad, died in his stateroom on board the PSNC's *Arequipa* on 14 April 1873 en route from Callao to Panama. Consul Murphy took charge of Forbes's remains and personal effects, notified his next of kin, and arranged for publication of an obituary notice in the *Callao and Lima Gazette*.[5]

A more sinister incident the following year also in-volved the PSNC's vessels and demanded the consul's at-tention. Charles Johnson, a naturalized American of Scot-tish birth, arrived in Paita in June of 1874 en route from Callao to Panama, accompanied by three other Americans, two of whom were apparently traveling under assumed names, the third a G.W. Blasdell. The quartet, whose pas-sage allowed them one stopover, checked into the Hotel Oriental in Paita. Sometime after Blasdell and the two mys-tery travelers departed on the next steamer, Johnson's body was found in a gully on the *tablazo*, near the road from Paita to Piura. For reasons that are unclear, shortly thereafter Blasdell returned voluntarily to Paita with the idea of trav-eling on to Piura. Before he could leave, however, he was arrested and jailed in connection with the Johnson murder, despite his protestations of innocence. Consul Murphy did what he could to protect Blasdell's interests and insure that he received a fair trial, but unfortunately the consular records bear no evidence of a resolution of this issue.[6]

As titillating as these two incidents involving *gringos* might have been for the residents of Paita, another scandal in which the United States consulate became deeply involved kept tongues wagging in northern Peru for years. On 30 October 1870 a naturalized American jewelry salesman and watchmaker named Joseph B.P. Liepsker arrived in Paita by steamer from Callao carrying jewelry reportedly valued at $20,000. On the passage from Callao, Leipsker was befriended by a Peruvian army officer, Major Carlos Varea Balta, who was commander of police for the Department of Piura and, coincidentally, a nephew of then-President José Balta.

In Paita, learning of Leipsker's plans to travel on to Piura on 4 November, Major Varea offered his services and those of his Afro-Peruvian aide-de-camp, a paroled murderer, as escort. After travelers found some of Leipsker's papers and other effects scattered about the *tablazo* along the road to Piura, the prefect of the Department of Piura ordered an investigation. Leipsker's hastily buried remains were found on 28 November and taken to Piura, where a forensic examination revealed that he had been shot.[7]

After receiving the news from the local authorities, Consul Columbus went to Piura to take charge of Leipsker's burial in the local cemetery and to claim his effects. He also asked the state department to notify Leipsker's widow in New York and directed two formal communications to the prefect, demanding a thorough investigation of the apparent murder. The local authorities refused to hand over the peddler's effects, claiming that they would be needed as evidence in the trial of his murderer.

After his trip to Piura, Dr. Columbus learned that Major Varea had escaped into southern Ecuador and that some of the jewelry stolen from Leipsker had been sold in Chiclayo and Lambayeque, coastal towns south of Paita. In

145

CHAPTER 11

February of the following year, Varea reappeared in Paita with the declared intention of picking up his mistress. She, however, having established a romantic liason with the sub-prefect in Paita, refused to accompany Varea. The major, in a fit of rage, went to the sub-prefect's office and shot him three times in front of witnesses. None of the wounds proved fatal. Major Varea fled out onto the *tablazo*, but was quickly captured and jailed in Paita.

Consul Murphy inherited the Leipsker case from Dr. Columbus when he took over the consulate on 1 April 1871. As a lawyer, he had a special interest in the legal proceedings against Varea, beyond the normal bounds of his consular responsibilities. Shortly after his arrival Murphy confided to Washington his doubts about the seriousness of the Peruvian judicial authorities, saying that "the murder of an American citizen is looked upon [here] as a matter of very little consequence."

The subsequent evasion of legal responsibilities at several levels of the Peruvian court system, where no one was eager to prosecute and sentence a nephew of the president, confirmed Murphy's misgivings. The judge of first instance in Paita refused to hear the case, on the grounds that the alleged murder of Leipsker did not occur within his jurisdiction. The judge of first instance in Piura, however, ordered him to open proceedings against Varea, which gave rise to an appeal to the superior court in Trujillo, which backed the judge in Piura. In the face of this setback, the judge in Paita appealed to the supreme court in Lima, which eventually agreed with the judge in Paita that the trial of Varea was not his responsibility. In order to break the impasse, a justice of the peace in Paita agreed to hear the charges against the major, but the results of this proceeding apparently had no validity.

The justice in Paita then sent the prisoner to Piura in irons. The restraint consisted of an S-shaped, 50-pound iron bar about three feet long called a *pletina*, which was placed around the prisoner's ankles and then hammered

closed to form a figure eight.[8] Shortly after the prisoner was
sent to Piura his mother - presumably President Balta's sister - arrived from Lima. After her appeals to the prefect to
remove the irons from her son's ankles were rejected, she
asked the judge of first instance to order their removal. The
judge agreed to have Varea examined by two doctors, who
found the prisoner to be in excellent health, and he too refused to order the removal of the *pletina*.

Family influence, however, soon prevailed over legal norms. The prefect in Piura was removed and the unco-
operative judge was given a leave of absence from his post.
In his absence, the justice of the peace in Piura ordered the
removal of the irons, in exchange for an alleged bribe of
$2,000 from the prisoner's mother. At the same time, an "in-
timate friend" of hers, a Juan Seminario, purchased the
strongest, fastest horse to be found in Piura. In mid-No-
vember 1871, almost a year after Leipsker's remains had
been found in a shallow grave on the *tablazo*, Varea escaped
from the jail in Piura with the help of the deputy commander of police, Lieutenant Luciano Carcelén, who
drugged the guard and provided the fugitive with an army
uniform. Varea again slipped over the border into southern Ecuador.

Some time later, according to an undated clipping
from the Panama City *Star and Herald*, the fugitive boarded
a steamer in the Ecuadorian port of Manta, enroute to
Panama. In the port of Buenaventura, a Colombian army
officer recognized Varea in a boot shop and tried to arrest
him, but the major eluded his would-be captor and disappeared into the dense jungle of the Colombian Pacific coast.
He was never heard from again. None of the jewelry fenced
in Chiclayo and Lambayeque was ever recovered. The ac-
commodating justice of the peace in Piura was arrested and
sent in irons to Trujillo, apparently the only actor involved
in the cold-blooded murder of the American jewel peddler
who was officially taken to account for his actions. In an
ironic twist of fate, however, eight months after Major

CHAPTER 11

Carlos Varea Balta escaped from the Piura jail with considerable help from family and friends, his uncle, President José Balta, was brutally murdered while being held prisoner in a cell in the San Francisco barracks in Lima.[9]

The murder of Joseph Leipsker in 1870 and the eventual failure of Peruvian justice to bring the assassin to account might be interpreted simply as a case of felonious cupidity exacerbated by cowardice and bureaucratic ineptitude. The fact, however, that Leipsker was a foreigner casts a slightly different light on the incident. In this light, the cold-blooded murder on the *tablazo* of a foreign merchant becomes a tragic sequel to the protectionist xenophobia that, according to Gootenberg, characterized the first decades of Peruvian independence and which periodically manifests itself even today.

12

Conclusions

What conclusions can be drawn from this history of the impact of the New England whaling industry on a remote Peruvian port of call in the nineteenth century and of the actors who played upon that stage?

A Different View of the New England Whaling Industry

For one thing, this study of the impact of the American whaling boom on Paita and of Paita's role as a socio-economic bridge between Peru and the United States in the nineteenth century provides a vision of the New England whaling industry from the other end of the telescope. It is not always a flattering picture.

It is clear that the men and boys who shipped aboard the Nantucket and New Bedford whalers, the flower of New England's youth, led a harsh life. They worked long hours under demanding physical conditions, and their reward was often uncommensurate with their sacrifices. Discipline on board the Yankee whalers was rigid and living conditions were spartan. The master was god, desertions were commonplace, and some crew members even burned their ships to get away. Unscrupulous ship captains, ever mindful of the bottom line - profit and loss - routinely evaded or violated United States laws designed to protect American seamen, regulate the whaling industry, and safeguard the Republic's good name abroad. Crew members were injured and became ill in faraway places. Many of them died. Those that survived were often crippled - physically and morally - for life.

The data available from consular returns also dramatically demonstrates the boom and bust nature of whaling, and the enormous potential for profit of a good cruise. The aggregate value of the oil cargoes on American ships calling at Paita in 1857 alone, for example, is equivalent to about $30,000.00 in 1991 dollars.[1] In contrast, the benefit accrued to Paita's economy from the goods and services provided to the whaling fleet was probably insignificant. But within the historical and spatial context of Paita it was important, for it provided the capital and the impetus for modern port facilities and improved communication, which in turn was the genesis of the expanding export economy from 1860 on.

The American Consular Hospital

This study also examines, for the first time, how American economic expansion overseas early in the nineteenth century - as exemplified by the whaling boom - was instrumental in the creation of an important expatriate American institution, the American Consular Hospital in Paita. This institution was an innovative response to a pressing social problem: how to provide adequate medical care to sick Americans far from home, men whose health and productivity were essential to the efficient operations of a key facet of the United States economy. The picture is far from complete because much of the documentation is missing. We know nothing about its capacity, how it was organized, or where it functioned. We do know that it was created in the late 1840s by official initiative to care for sick American seamen, when it became apparent that no local institution was satisfactory. We also know that from the beginning it was administered and staffed by expatriate professionals, and that it was subsidized by its principal beneficiary, the whaling industry, and by the United States government. The roles of the American consul and the director of the American hospital were intertwined throughout its 20-year history, giving rise to confusion and misunderstanding, and leading in the end to breaches of faith and serious charges of malfeasance.

It is probably no coincidence that of the five men who served significant periods of time as American consul in Paita - Ruden, Ringgold, Winslow, Columbus, and Murphy - all but the first and last were medical doctors. Nor was it mere happenstance that during this period American consuls serving in neighboring Tumbes and Lambayeque - Oakford and Montjoy - were also physicians. What accounts for this concentration of American medical professionals in northern Peru in the middle of the nineteenth century?

In the case of Paita, the simple answer is that they were attracted there by the existence of an expatriate American hospital. This explanation also applies to the case of Dr. Thomas Oakford, who was the first American physician of record at the Consular Hospital in Paita, and only later became consul in Tumbes. But the facile answer often begs the broader question: why did American doctors, graduates of prestigious universities in some cases, find it necessary to emigrate to Peru in order to find remunerative, professionally satisfying work?

The answer to this more fundamental question lies in the status of medical education in mid-century North America and the dynamics of the domestic job market for qualified physicians. Beginning in the 1840s, when the individual states began to abolish licensing laws that regu-

lated the professions, there was a proliferation of propri-
etary medical schools and colleges, some serious, many not
so serious, that turned out large numbers of physicians.
From the first decade of the nineteenth century to the 1850s,
the supply of medical graduates increased over 50 times.
Until the outbreak of the Civil War, however, the supply of
medical doctors in the United States surpassed the demand,
forcing many young medical school graduates to seek their
fortunes on foreign shores.[2] Thus, along with American
merchants, sailors, engineers, and mechanics, American
doctors working in Peru were part of the successful North
American challenge to earlier British economic and social
hegemony in the former Spanish colonies. And because of
their youth, like many other expatriate Americans in nineteenth-century Peru, many of these young physicians fell
prey to the seductive charms of the *criolla peruana*, marry-
ing and putting down deep roots in Peruvian soil.

   The documentary record does not allow me to hazard a guess as to the American Consular Hospital's patient
recovery rate. We simply do not know how many seamen
were cured and returned to duty, or how this success rate
compares with that of contemporary institutions elsewhere,
either in the United States or Latin America. The fact that
almost all of the complaints extant in the consular records
are about the financial management of the institution, rather
than about its physical conditions or mortality rate, suggests that it probably did a fairly good job. Only when it
became a financial burden to the Federal government, during the Civil War, and when its reason for being was diminished as a result of the dramatic decline in whaling associated with that conflict as well as and with the reduced
demand for sperm-whale products, did it become necessary and possible to close it. Simultaneously, significant improvements in transportation made it possible to send sick
seamen home via the Isthmus of Panama without running
too great a risk that they would die en route.

CONCLUSIONS

The Economic Impact of the Whaling Boom on Paita

Paita is an unimportant place today, a small provincial port, a backwater. Despite the sustained growth of Piura cotton production up until World War I, which was exported through Paita, by the second decade of this century the port's earlier economic vitality had waned considerably. In 1911 an American visitor remarked that Paita was "at one time a place of great commercial importance." He noted the port's former role as a trading center, but did not even mention the impact of the New England whaling industry on northern Peru.[3] But we know that Paita was not always so unimportant. Despite its distance from Lima and the core of economic and political power, Paita was important to Peru's economic growth and development from the first years of the conquest until the end of the nineteenth century. As we have seen, one-third of Peru's customs revenues were derived from Paita at the end of the last century.

Paita's *espacio económico* always generated products for which there was a significant demand along the Pacific coast, in the United States, and in Europe. Moreover, during the colonial period the port of Paita played a well-defined, essential economic role as a provisioning place for sailing ships and as an entrepôt. Its *raison de'etre* did not diminish with the advent of independence. The technological and commercial innovation that would eventually make Paita less important as a port of call - the advent of steam navigation on the Pacific coast of South America - did not impact on Paita until the 1840s. The introduction of steamships meant that the coastal trade was no longer subject to the vagaries of the Humboldt current, and diminished Paita's importance as a port of call and an entrepôt. But the high cost of transporting goods and passengers via steamers that prevailed throughout the nineteenth century ensured the survival of the competition - traditional sailing vessels - and helped to sustain Paita's economic vitality.

CHAPTER 12

Sooner or later Paita was condemned to wane. But the arrival of the American whaling ships in the early 1830s prolonged its economic usefulness by a third of a century. The whaling boom brought money to Paita; it helped to stimulate the construction of port facilities, the introduction of urban amenities, and improvements in land transportation. And it helped to lay the groundwork for the dramatic expansion of the export economy during the European "cotton famine" of the 1860s and the dawn of the age of petroleum on the Talara peninsula late in the century.

The whaling boom had its ups and downs, but from the early 1830s until 1865 this American presence in Paita helped to sustain and expand an already diversified, well-articulated, peasant-based service economy. The service sector that catered to the whalers - the "small trade" in contemporary terms - embraced many Peruvians, ranging from wealthy merchants and professionals to farmers, muleteers, water carriers, seamen, ship carpenters, and firewood gatherers. The wealth generated by whaling flowed into the Paita economy in a myriad of ways: in the purchase of food and supplies, in the pursuit of entertainment, and in the provision of services, including that service for which seaports are most infamous. Was this wealth invested or consumed? Did it benefit only private individuals or did the community at large profit from it? Did it end up in the hands of the few, or did it make a difference in the economies of many modest families? The answers to these questions, if they exist, await further investigation. But it seems clear that the economic impact of the whaling boom on Paita was a key factor in conditioning the port and its economic hinterland to assume the important role that it later played in the agricultural export economy that became the norm for much of Latin America in the second half of the nineteenth century.

CONCLUSIONS
The Social Impact of Whaling on Paita

If the whaling boom's economic impact on Paita is difficult to quantify and to assess, so too is its social impact. In addition to the many officers and crew members of the New England whalers who landed there, spent their money there, sought sexual gratification there, and perhaps even established consensual unions with Paita women that may have resulted in progeny, the boom attracted a large and diverse group of foreign "landlubbers" to Paita. They were Americans and others who came to seek their fortune and departed, and others who cast their lot with Paita and Peru. They were pioneers and entrepreneurs, they were merchants, diplomats, and scoundrels. They came from very different backgrounds and expected and got different things from Paita. Some were successful and some were failures, but few of them would have even heard of Paita, let alone chosen to live there, had it not been for the whaling boom.

This study provides a better picture of their backgrounds, their role in the growth and development of the Paita hinterland, and their contribution to its diversity, than has been available to date for any similar peripheral region of Latin America. But they were not unique. Their counterparts can be found all over the hemisphere in the nineteenth century. What we have learned about the foreign community in Paita, however, may enable us to develop some useful prototypes for expatriate communities in Latin America, about which relatively little is known.

Prototype One
The Middling Trader/Merchant

Of modest economic and social background with few connections beyond his own family, this prototypical individual was a first-generation American or perhaps a recent immigrant to the United States, but was thoroughly American in outlook and identified completely with the political

155

and economic ideals of the young, dynamic North American republic. He was educated in trade and in the school of experience and hard knocks. In Paita he sought a niche for himself, a place where he could use his trading skills and his market experience to make his fortune, Horatio Alger-style. He often put down roots in the community and, since he left behind him little in the way of material goods, his ties with the United States became attenuated. Only modestly successful and without friends in high places, this prototype sought a diplomatic appointment as protection against the xenophobic nationalism of the early republican elites, the economic vagaries of the mercantile world, and the arbitrariness of local politicians. Alexander Ruden, the first American consul, and his English clerk Gerard Garland might fall into this category. So too might Dr. Rafael Columbus and John Murphy of Brooklyn.

Prototype Two
The Yankee Entrepreneur Adventurer

This prototypical individual came from a background of considerable local distinction and was heir to a maritime tradition that extolled and romanticized travel and residence in exotic places as a way of making one's mark in the world. Young, fairly well-educated and unfettered by family responsibilities, he was willing to take economic risks such as investing in a cotton plantation in Peru in the 1830s, because he had something to return to, a socioeconomic position in his community of origin, a distinguished family name, a support group, and a chance to start over. Pioneer cotton planter Frederick Dorr clearly fits this category.

CONCLUSIONS

Prototype Three
The Elite But Impoverished Professional

This prototypical individual was often the youngest son in a large family with little hope of inheriting anything but a distinguished surname. Growing up, he had the advantages of class, political connections, a quality education, and the remnants of the family fortune which at least gave him a marketable skill and the freedom from want to allow him to indulge a thirst for travel and adventure. With no estates to inherit, no position in the community to assume, and no family business waiting for him, he was free to test his skills in exotic, faraway places like South America. There he came face to face with the need to earn a living, and used his family's social and political connections to secure a diplomatic sinecure that gave him both status among the local elite - his natural peers - and a degree of economic security. His youth and sophistication helped to win him entree into the circles of the local elite, to appreciate Peru's rich but very different cultural tradition, and perhaps to find a mate. Representative of this type is Fayette Ringgold, the second consul in Paita, and perhaps his colleague Dr. Thomas Oakford.

Prototype Four
The Engineer

Treated with the respect and awe reserved today for astronauts, the prototypical engineer was lured to Peru by the challenge of bringing modernity to a backward but potentially rich part of the world. His dream was to cast down the barriers to trade and communication, to build a railroad where none existed before, to open an area of untapped natural resources, to build an aqueduct or an irrigation project, in short to usher in "progress." The prototypical engineer soon realized, however, that economic problems, cultural inertia, and politics were formidable obstacles to

157

progress. He often channeled his energies and skills into private, speculative activities with the potential to make him a very wealthy man. Typical of this stereotype were the Baltimore engineer Alfred Duvall and his contemporary William Stirling.

Prototype Five
The Dilettante Intellectual

The nineteenth-century dilettante intellectual was lured to South America by adventure and the prospect of undiscovered frontiers - both physical and intellectual. He was driven by an incurable wanderlust, willing to live in remote and uncomfortable places without thought of material reward. His was a lifetime quest. His education and experience intensified his thirst for knowledge but, unless he had abundant private fortune, from time to time he was reduced to using his intellectual and political connections to secure a government position or a diplomatic sinecure - the nineteenth-century equivalent of a research grant - as a temporary economic expedient and perhaps an entree into the elite or intellectual spheres of the community. Dr. Winslow and the British botanist Richard Spruce spring to mind as typical of this kind of individual.

Prototype Six
The Outcast and the Misfit

The most anonymous of these prototypes, the outcast or misfit who washed ashore in a place like Paita in the nineteenth century, might have been a poor boy with a wanderlust, a renegade, a social misfit at home or a fugitive from justice. The dangers and uncertainties of a mariner's life may not have seemed so bad when compared with the lack of opportunity at home. But life on a whaler hardened and corrupted many an innocent boy, and killed a few as well. The outcast did not choose to come to Paita, it just happened. Once there he drank, stole, gambled, and fought, becoming a problem for Peruvian and consular officials alike. Unfortunately, the story of Paita's role as a socio-economic bridge between Peru and the United States also has its share of this kind of character, its "One Arm Bills" and its Captain Hillmans.

What did these expatriate Americans and foreigners of other nationalities contribute to Paita? They brought a different world view, different values, a different work ethic, and different ideas about prestige and success. Some were crass seekers after wealth with no scruples. Others adapted more graciously to the ethos of their adoptive homes. Some were harsh critics of the customs and the ethics of Peruvian society. Many were frustrated by the Peruvians' failure to respond to the same stimuli and incentives that they responded to, to imbibe the entrepreneurial spirit, to equate the pursuit of personal advancement with the quest for "progress."

Some foreigners brought technical skills that were very much needed and appreciated in Paita, the ability to cure or comfort, the knowledge of laborsaving devices and the experience to build systems that generated wealth and enhanced the quality of everyday life. Others brought mercantile skills and familiarity with foreign markets, a knowledge or sense of what was saleable and what was not, the experience needed to deal successfully with both suppliers and consumers, and the personal integrity needed to sustain relationships of confidence and trust.

These expatriates often provided the capital needed to undertake a commercial venture, to advance credit to suppliers, to take advantage of opportunities, and to invest in speculative enterprises like cotton planting.

What did they get from Paita? A chance to catch the brass ring, to strike it rich, to become a part of the local elite, or to go home a success. They satisfied their curiosity about the world around them and learned something about themselves in the process. And the really successful ones developed the language and cultural skills to deal success-fully in a very different milieu, a sensitivity and under-standing of the host culture and society that bred respect and admiration and inspired the same.

As is generally the case, this study of the role of Paita as a bridge between Peru and the United States in the nineteenth century raises as many questions as it attempts to answer. I would like to be able, for example, to shed more light on the later careers of Alexander Ruden and Fayette Ringgold, the first two American consuls in Paita, but my sources are mute. Given their different backgrounds, their expectations in going to Paita and their standards for mea-suring "success" in life were probably very different.

I would also like to learn more about the back-ground of the first American doctor at the Consular Hospi-tal in Paita and subsequently American consul at Tumbes, Dr. Thomas Oakford, and about his colleague in Lambayeque, the enigmatic Santiago Coke Montjoy, but the otherwise rich resources of the Daughters of the American Revolution Library in Washington were little help.

My modest success in reconstructing the family background of Frederick Dorr, the Massachusetts cotton planter, is offset by my frustration in being unable to learn how successful he was in Peru, why he returned to the United States, and what he did with the rest of his life. The cases of Fayette Ringgold, and Frederick Dorr, both mem-bers of old and distinguished families and both raised in an environment of privilege, raise the questions, "What lured them to South America, at a time when travel was arduous and difficult and the environment often less than bucolic, and what kept them there?"

The motives of Dr. Charles Winslow and the Balti-more engineer Alfred Duvall for going to Paita are some-what clearer. Winslow was infected with an incurable wan-derlust and an unquenchable thirst for knowledge, while at the same time being blessed with a modest personal fortune that allowed him to indulge those passions. Duvall was a problem solver.

Going beyond personalities, one is led to broader but equally unresolved mysteries. How lucrative really was the New England whaling industry during its apogee, and how much of that wealth was transferred - through the provision of goods and services - to Paita and its hinterland? Was this hypothetical wealth widely distributed among the primary providers of these goods and services, or did established land tenure patterns and limits on access to capital and influence result in a concentration of this wealth in a few hands, either Peruvian or foreign? Did wealth help to create a significant class of small freeholders who were able to defend themselves economically against the encroachments of export-oriented plantation agriculture, or were the conditions that made this possible already in existence in the Piura region? Is it possible to clearly establish a causal relationship between the American presence in Paita and the onset of modernization in northern Peru? Was the ability of Paita's *espacio económico* to respond with alacrity to the opportunities afforded by the "cotton famine" of the 1860s the direct result of the intense American economic and social presence in Paita between 1830 and 1865, or would it have happened even if the New England whalers had not rendezvoused in Paita for over a third of a century?

Outpost of Another Empire

Despite the unanswered questions, I hope that this study contributes to the growing understanding of the important roles played by peripheral areas in the economic and social life of nineteenth-century Latin America. The case of Paita clearly shows that despite its distance from the focus of economic and political power, it was not a jerkwater place where everyone lolled in hammocks and played the guitar all day. Quite the contrary: from the days of the conquest Paita was an essential outpost of the Spanish empire in Peru, a key player in the viceroyalty's economic life and structure. Paita was also a key player in Peruvian economic life in the early nineteenth century. It was a channel for the products of the northern Peruvian hinterland and it was a service economy for the New England whalers. By the 1860s, however, thanks at least in part to the economic and social impact of the Yankee whalers and the fortuitous creation of a lucrative market for Peruvian cotton and petroleum in Europe, Paita was transformed into the outpost of another kind of empire - the North Atlantic empire of foreign investment in the production and export of raw materials for the industrialized world.

Appendix

Arrivals of American Whaling and Merchant Ships at Paita, 1832-1865

Name	Home Port	Tonnage	Year
Abigail	New Bedford	309	1832,40,44,45
Active	New Bedford	333	1857,58,59,62
Acushnet		351	1858
Adelaide		373	1851
Advance		274	1865
Afton	New Bedford	249	1860
Alexander	New Bedford	421	1843
Alfred Gibbs	New Bedford	425	1860,62
Alfred Tyler		225	1849
Almira	Edgartown	362	1835,48

Name	Home Port	Tonnage	Year
Alpha	Nantucket	345	1835
Alto	Fairhaven	197	1841
Alto	New Bedford	236	1860
Amazon	Fairhaven	318	1842,43,44
American	Sag Harbor	284	1847
American	Nantucket	340	1839,40,42,43,44,46,47
Amethyst	New Bedford	359	1848,49,51
Anaconda	New Bedford	383	1862
Anglona		81	1853
Ann Alexander	New Bedford	252	1835,40,42,49
Ann McKim	Baltimore	494	1848
Anne	Bristol	222	1835,39,46
Ansel Gibbs	Fairhaven	359	1857,58
Arab	Fairhaven	275	1832
Ardennes	San Francisco	231	1851
Arnolda	New Bedford	360	1847,60,62
Ascutna		429	1851
Atkin Adams	Fairhaven	330	1849,57,60
Atlantic	New Bedford	321	1859
Audley Clarke	Newport	331	1835,42
Augusta	New Bedford	344	1835,39,40
Augusta Buckhorn		148	1845
Aurora	Nantucket	345	1832,35,40,41,43,47
Aurora	Westport	351	1857,59
Awashonks	Falmouth	355	1841,42
Bartho. Gosnold	New Bedford	356	1835
Balaena	New Bedford	301	1832,42,46-48,57,59,60,62
Balance	Bristol	321	1835
Baltic	Nantucket	409	1842
Benj. Cummings	Dartmouth	391	1858
Betsy Williams	Stonington	399	1857
Boy	Warren	251	1844,46
Boy		359	1842
Braganza	New Bedford	469	1835
Brandt	New Bedford	301	1842
Brighton	New Bedford	354	1832
Brutus	Warren	470	1852
Burnham		248	1851
Cairses		327	1857
Callao	New Bedford	323	1844,46

Name	Home Port	Tonnage	Year
Cambria	New Bedford	362	1832,39,40
Canada		270	1840,43,46,47,52,53,58,60
Canada	New Bedford	545	1848
Canton	New Bedford	409	1832,35,40,49
Cape Horn Pigeon	Dartmouth	300	1858,59,60
Carib	San Francisco	205	1852
Carmelita		181	
Caroline	New Bedford	364	1848
Catalpa	New Bedford	260	1857,59
Catawba		270	1850
Catawba	Nantucket	335	1842, 46, 54
Catawba		296	1848
Cachelot	Mattapoisett	230	1848,59,60
Catharine	New London	384	1842
Catherwood	Westport	199	1854
Cayuga		246	1848
Ceres	Wilmington	328	1843
Champion	Westport	209	1851,55
Champion	Edgartown	399	1840
Chariot	Warren	359	1842
Charles	New Bedford	290	1839,44
Charles & Edward	Dartmouth	150	1864
Charles & Henry	Nantucket	336	1841
Charles Carroll	San Francisco	376	1849
Charles Drew	New Bedford	344	1835
Chas. W. Morgan	New Bedford	351	1842,43,44
Charlotte	Sag Harbor	230	1852,53
Chassan		113	1852
Chelsea	New London	396	1835
Cherokee	New Bedford	261	1841
Chili	New Bedford	291	1851,57,58
China	New Bedford	370	1847,60,62
Christ. Mitchell	New Bedford	387	1835,40,42
Civilian		161	
Clara Bell	Mattapoisett	295	1859,60,63
Clarice	New Bedford	237	1839,41,60
Clifford Wayne	Fairhaven	304	1857,58,59
Columbia	Nantucket	329	1849
Columbus	Fairhaven	382	1842,48,49,52,60
Columbus	Nantucket	344	1842

APPENDIX

Name	Home Port	Tonnage	Year
Commo. Morris	Falmouth	355	1860,62
Commshubrick		62	1848
Congaree	New Bedford	321	1847,48,49,51-54,57,58,62
Congress	New Bedford	339	1832,40,42
Congress	New Bedford	376	1847,49
Constitution	Nantucket	318	1840,41,44,45,49,53
Constitution	Nantucket	400	1862
Cora	New Bedford	220	1835,42
Coral	New Bedford	370	1841,44,47,48,49
Corea	New London	365	1860
Corinthian	New Bedford	401	1832,40,44,45,46,63
Cornelia	New Bedford	291	1842
Corsair	New York	160	1840
Cortes	New Bedford	382	1840,42
Courier	Dorchester	293	1853
Courier	New Bedford	381	1857
Covington	Warren	351	1848
Cristobal Colon		340	1849,50
Damariscotta		102	1851
Darling		670	1860
Dartmouth	New Bedford	336	1841,42,59,60
David Paddack	Nantucket	352	1842,43,44,45,46
Desdemona	New Bedford	295	1840,42,47,48,50
Dodge		274	1860
Dominga	New Bedford	230	1857,59,60
Dragon	New Bedford	307	1832
Dromo	Warren	267	1851,64
Dryade	New Bedford	262	1848,50
Dumbarton	New Bedford	199	1850
E.A. Luce	Edgartown	132	1858
Eagle	Nantucket	336	1839,42,45,46,47,48
Edward	Hudson	274	1847,48,58,59
Edward Carey	Nantucket	350	1851,53,63
Edward Fletcher		261	
Edward Grinnel		388	1835
Edward Lacey			1860
Elizabeth Swift	New Bedford	425	1860
Eliza	Salem	260	1842
Eliza	San Francisco	279	1852
Eliza Adams	Fairhaven	403	1840,41

166

AMERICAN SHIPS CALLING AT PAITA, 1832-1865

Name	Home Port	Tonnage	Year
Elizabeth			1860
Elizabeth	Salem	398	1845,47
Elizabeth Ann		370	1849
Elizabeth Starbuck	Nantucket	381	1840
Eben Dodge	New Bedford	221	1859
Emerald	New Bedford	359	1840,48
Emily	New Bedford	294	1860
Emily Morgan	New Bedford	370	1835
Emma	New Bedford	245	1853
Emma C. Jones	New Bedford	346	1863
Empire	Nantucket	403	1848,51
Enterprise	New Bedford	291	1835,40,41
Equator	Nantucket	262	1835,44,47
Erastus Corning	New Bedford	225	1862,63,64,65
Erator		196	1850
Ermina Jane		1096	1858
Essex	Bristol	200	1835
Eugenia	New Bedford	356	1852,60
Euphrasia	Baltimore	487	1846
Euphrates	New Bedford	364	1839,40
Excell		81	1854
F.A. Everett		241	1850
Fabius	Nantucket	432	1840,42,43
Fellowes	Stonington	268	1846
Finders		193	1835
Floyd		223	1850
Formosa	New Bedford	450	1845
Forrester	Dartmouth	243	1832,35,40
Foster	Nantucket	317	1832
Frances	New Bedford	348	1835,41,48
Frances Henrietta	New Bedford	407	1840,41
Franklin		227	1858,59,60
Franklin		235	1847
Franklin	Nantucket	246	1839,43,46,48
Franklin	New Bedford	273	1852,59,60
Franklin	New Bedford	333	1848
Franklin II	New Bedford	218	1844,45,46,51,60
Freemont	New Bedford	273	1850
Friendship	New Bedford	366	1832,42
Ganges	Nantucket	315	1842

APPENDIX

Name	Home Port	Tonnage	Year
Ganges	Bristol	380	1832
Garland	New Bedford	243	1843,44,57
Gazelle	Nantucket	340	1853
General Cobb		121	1850
General Jackson	Bristol	329	1832
General Pike	New Bedford	313	1860
General Scott	Fairhaven	333	1840,44,45,46,48,57,59
General Wool		369	1852
Genge		360	1843
George	New London	290	1858
George	Fairhaven	359	1841,42
George and Mary	New London	357	1832
George and Susan		256	1853
George and Susan	New Bedford	356	1835,40,42,49
George Clinton	Hudson	425	1835
George Henry	New London	303	1840,41,43,44,46-51,57,58
George Howland	New Bedford	374	1840,42,44,45,47-49,58
George Porter	Nantucket	285	1841,44,46
Geo. Washington	Wareham	373	1848
Gideon Howland	New Bedford	380	1832
Glide		169	1847
Go Ahead		410	1864
Golconda		240	1860
Golconda	New Bedford	331	1835,40-42,44,45,48,49,57,58,62
Gold Hunter	Fall River	280	1857
Governor Fenner	Bristol	375	1832,35
Grace Sargent	Yarmouth	950	1865
Gratitude	New Bedford	337	1832
Greyhound	Westport	249	1859,60
Hanson		370	1842
Harvest	New Bedford	262	1858
Harvest	Nantucket	360	1850,51
Hecla	Sippican	207	1850,60,63,64,65
Hector	New Bedford	380	1832,35,39,40,41,43,62,64
Helen M.		84	1850
Helen Mar	New Bedford	367	1857,58,59
Helper		261	1846
Henry	Sag Harbor	346	1841,42,45,52
Henry Astor	Nantucket	375	1832,40
Henry Clay	Nantucket	385	1844,45,49

AMERICAN SHIPS CALLING AT PAITA, 1832-1865

Name	Home Port	Tonnage	Year
Henry H. Crapo		199	1853
Henry Taber	New Bedford	355	1860,62,63
Herald	New Bedford	274	1843,59,60
Hercules		371	1851
Hermoine		259	1850
Hero	Nantucket	313	1835,42,48
Heroine	Fairhaven	337	1848
Hesper	Fairhaven	262	1846,47,49,51,57,60,62,63
Highlander	Mystic	238	1848
Hobomok	Falmouth	414	1846
Hope	New Bedford	295	1848
Hortensia		263	1845
Hula		207	1858
Huron	Sag Harbor	292	1832
Hydaspe	New Bedford	313	1846,49,57,58,60,62
Independence	Nantucket	311	1832
Independence	New Bedford	318	1832
Inez	Newbury	699	1853
Iris	New Bedford	311	1832,41,44
Isaac Coffin		338	1832
Isaac Howland	New Bedford	399	1832,39,40,41,46
Isabella	New Bedford	410	1849
Islander	Nantucket	347	1858
Izaak Walton	New London	437	1849
J.M. Ryerson		177	1850
James Allen	New Bedford	353	1846
James Caskie	Newbury	283	1852
James Loper	Nantucket	348	1857,59,60
James Monroe	Fairhaven	424	1841,42,46
Jane	Somerset	231	1844
Janet	Westport	194	1849
Janet Ann			1860
Jason	New London	335	1832
Java	New Bedford	295	1857
Jeannette	New Bedford	340	1843
Jenny Elizabeth		380	1844
John	New Bedford	308	1841,42
John A. Parker	New Bedford	342	1858,59,60
John A. Robb	Fairhaven	273	1846
John Adams	New Bedford	268	1845,46

APPENDIX

Name	Home Port	Tonnage	Year
John Adams	Nantucket	296	1842,43,44,47,48
John and Edward	New Bedford	318	1846
John Howland	New Bedford	377	1848
John R. Gardner		190	1849,50
Joseph Grinnell	New Bedford	458	1859,60
Joseph Maxwell	Fairhaven	301	1841,42,44,46
Joseph Meigs	New Bedford	356	1853,59,60
Joseph Starbuck	Nantucket	416	1840
Joshua Bragdon	New Bedford	270	1858
Juliet		263	1849
Juliet		275	1851
Junior	New Bedford	377	1842
Kaluna		149	1854
Kingston	Nantucket	312	1842,43,49
Kirkwood	Nantucket	201	1850
Laetitia	New Bedford	274	1858,59,63
Lafayette	New Bedford	260	1849
Lafayette	New Bedford	311	1849,53,57,58,60
Lafayette	Warren	341	1851
LaGrange	New Bedford	280	1858,59,60
La Mar	Baltimore	326	1854,57,58
Lancer	New Bedford	396	1857,58,59,62
Laura	Plymouth	219	1852
Laurens		624	1853
Leader	New Bedford	200	1832
Leland	Boston	347	1845
Leo		136	1851
Leonesa		290	1859
Leonidas	New Bedford	231	1851, 52
Leonidas	Fairhaven	353	1832
Levant	Wareham	219	1843
Lexington	Nantucket	398	1841
Lima	Nantucket	286	1832,35
Lion	Providence	298	1842
Loan	Edgartown	262	1835
Logan	New Bedford	302	1832,35,40
London Packet	Fairhaven	335	1840,49
Long Island	Brookhaven	174	1852
Loper	Nantucket	317	1832
Louisa	Baltimore	316	1844

AMERICAN SHIPS CALLING AT PAITA, 1832-1865

Name	Home Port	Tonnage	Year
Louisiana	New Bedford	293	1860,65
Lydia	Nantucket	351	1841
Lydia	Fairhaven	358	1849
Magnet	Warren	355	1840
Magnolia	New Bedford	396	1839,40,41
Magrut		355	1832
Mallory		291	1850
Marcus	Fairhaven	286	1841,42,43
Margaret		246	1850,51
Margaret	Newport	375	1839,43,44
Margaret Scott	New Bedford	307	1842
Maria	Nantucket	365	1835,44,51,52,53
Maria	New Bedford	202	1832
Maria Theresa	New Bedford	330	1832
Mariner	Nantucket	348	1850,53
Mars		207	
Mary & Susan	New Bedford	407	1862,63
Mary Adeline		185	1852
Mary Ann	Mattapoisett	214	1860
Mary Ann	Fairhaven	335	1844,47,48,57
Mary Frances	Warren	311	1849,59,60
Mary Frazier	New Bedford	288	1843
Mary Mitchell	Nantucket	354	1845
Mary Sensa		252	1849
Mary Wilder	New Bedford	213	1853,57,58,59,63
Massachusetts	New Bedford	364	1841,43
Matilda Sears	Dartmouth	300	1858,59
Matthew Luce	New Bedford	410	1860
Maury		394	1841
Mechanic	Newport	335	1835,43,44
Memphis	New York	798	1860
Menkar	Newport	370	1849
Mercator	New Bedford	246	1854
Mercury	New Bedford	339	1832,41,47,50
Meridian	Edgartown	381	1835
Merlin	New Bedford	347	1857
Mermaid		326	1858
Messenger	New Bedford	291	1845,46,48,49
Metacom	New Bedford	360	1842
Michel Angelo	Boston	783	1852

APPENDIX

Name	Home Port	Tonnage	Year
Midas	New Bedford	326	1840
Midas		362	1839
Milton	New Bedford	388	1832
Minerva	New Bedford	195	1832
Minerva	New Bedford	407	1835
Minerva Smyth	New Bedford	335	1841,42,49,50
Missenquis		277	1832
Mobile	New Bedford	263	1835,45,46
Monterrey		78	1859
Montezuma		253	1845,51,52
Montgomery	Nantucket	248	1847
Montgomery	New Bedford	248	1853,59,60
Morning Star	New Bedford	305	1858,59,60
Mount Vernon			1840
Mount Wollaston	Fairhaven	325	1842
Mousam	Mousam	221	1851,52
Nantucket	Nantucket	350	1842,48,49,53
Napoleon	Nantucket	360	1845
Napoleon	New Bedford	360	
1857			
Nautilus	New Bedford	340	1832,43
Nautilus	New Bedford	374	1853,57,58,60
Navigator		333	1852
New Bedford	New Bedford	351	1841,44,46,49,50
New England	Poughkeepsie	375	1841
Niantic	Warren	451	1849
Niger	New Bedford	437	1845,49,62,63,64
Niger		476	1853
Nile	New Bedford	321	1841,45
Noritoa		340	1846
Norman	Nantucket	338	1847,49,50,62,63
North America	New London	388	1832
Nye	New Bedford	211	1835,45,46
Obed Mitchell	New Bedford	354	1842
Ocean	New Bedford	350	1841,51,60,64
Octavia	New Bedford	257	1832
Ohio	New Bedford	237	1853,57,60
Ohio	New Bedford	383	1863
Omega	Nantucket	363	1832,41
Ontario	Nantucket	354	1844,48,51

AMERICAN SHIPS CALLING AT PAITA, 1832-1865

Name	Home Port	Tonnage	Year
Ontario		199	1844
Orbit		171	1850
Orbit	Nantucket	351	1835,40,41
Oregon	Fairhaven	339	1846,47,48
Orion	Nantucket	354	1839
Orpheus		573	1842,44
Osceola III	New Bedford	200	1860,63
Ospray	New Bedford	169	1832
Osprey	New Bedford	236	1858,60
Pacific		200	1865
Pacific	New Bedford	314	1860
Pacific		323	1846
Pacific	New Bedford	385	1832,35,40,41,42,43,45,50
Page		149	1850,51,52,62
Page		414	1860
Palmetto		282	1859,65
Pantheon	New Bedford	284	1848
Parker	New Bedford	406	1840,41
Patuxent		95	1848
Paulina	New Bedford	271	1854
Pavilion		150	1850,49,51,52
Perseverance		103	1851
Persia	New Bedford	240	1848,50,51
Peru	Nantucket	257	1849
Peruvian	Nantucket	334	1841
Phebe	Nantucket	379	1835,40
Phenix	New Bedford	423	1857
Phillipe De La Noye	Fairhaven	383	1858
Phoenix		303	1832
Phoenix	Nantucket	323	1839,41,42,46
Pilgrim		120	1850,52,58
Platina	Westport	266	1848,49,52
Pocahontas	New Bedford	341	1832
Pontiac		69	1850,51,52
Potomac	Nantucket	356	1842
President	Nantucket	293	1832,39,41,45,48,53
President	New Bedford	293	1857,58,60
Providence	New York	71	1851,52
Quickstep		338	1852
Quickstep		350	1851

APPENDIX

Name	Home Port	Tonnage	Year
Rajah	New Bedford	250	1832
Rambler		318	1832,50
Rebecca Sims	New Bedford	400	1851
Rhine	New Bedford	174	1849
Rhone	New York	471	1844
Richard Mitchell	Nantucket	386	1832,41,42
Richmond	New Bedford	291	1832
Rio		221	
Roanoke	New York	318	1845
Robert Edwards	New Bedford	356	1832,43,44,47
Roman	New Bedford	371	1841, 48,50
Rosalie	Warren	323	1840,45,48,49
Roscius	New Bedford	301	1857
Roscoe	New Bedford	362	1835,44,45,49,62
Rose	Nantucket	350	1848
Rousseau	New Bedford	305	1835,40,46
Russell	New Bedford	301	1832,48
Russell		348	1851
Russell	Dartmouth	387	1842,43,44
S. Tarbox		549	1863
Sacramento	New York	187	1851
Sacramento	Westport	218	1859,60,62
Saint Joseph	Baltimore	305	1847,50
St. Mary		108	1853
Samuel & Thomas	New Bedford	191	1865
Samuel Adams		274	1846
Samuel L. Southard	New York	176	1844
Samuel Thomas		190	1865
Samuel Thomas		140	1860
Santiago		99	1854
Sapphire			1840
Sappho	New Bedford	319	1849,50,52,53,54,57,58
Sara		291	1847,50,51
Sara Rourk		323	1841,42
Sarah		281	1850
Sarah	Warren	286	1845,47,48
Sarah	Nantucket	495	1841,42
Sarah Frances	Fairhaven	301	1845,46
Sea Queen	Westport	263	1853,59
Seaman		240	1843

AMERICAN SHIPS CALLING AT PAITA, 1832-1865

Name	Home Port	Tonnage	Year
Seaman		244	1846
Seconet	New Bedford	400	1857,58
Seine	New Bedford	281	1863,64
Serene		313	1840
Sharon	Fairhaven	354	1840,49
Smyrna	New Bedford	219	1842,47,48,60
Sonia		230	1859
Spartan	Nantucket	333	1840,44-46,48,49,57
Splendid	Edgartown	392	1840
Spray		320	1850, 52
Stafford	New Bedford	206	1853
Statira	New Bedford	346	1841
Stella	New Bedford	338	1857,62
Stephania	New Bedford	315	1845
Superior	New Bedford	276	1852,53
Susan	Nantucket	349	1848,49
Susan		317	1832
Swift	New Bedford	321	1846
Swift		390	1848
Thomas Pope	New Bedford	323	1858
Three Brothers	Nantucket	384	1842,49
Thule	Nantucket	286	1843
Timoleon	New Bedford	346	1832
Trident	New Bedford	448	1832, 63
Triton	New Bedford	232	1835,42,44,58,59
Tweed		306	1843
Two Brothers	New Bedford	288	1853
Tyrone		537	1851
Uncas	Falmouth	412	1832
United States	Westport	217	1841
Valparaiso	New Bedford	402	1849
Velasco		271	1850
Venezuela	San Francisco	195	1853
Vigilant	New Bedford	282	1853,60,62
Vineyard	Edgartown	380	1849
Virginia		346	1842,44,49,57
W.A. Farlton		75	1851
W.B. Andrews		303	1857,59
W.L. Archer		321	1843,58,59
W. Rock		280	1832

APPENDIX

Name	Home Port	Tonnage	Year
Walter Scott	Nantucket	339	1845,50
Washington		236	1835
Washington		239	1835
Washington	Nantucket	308	1840,41,42,45
Wave	New Bedford	197	1857
Waverly	New Bedford	327	1840
Weymouth	Nantucket	330	1832
Whig		221	1843
William & Eliza	New Bedford	321	1848,49,51
William & Henry	Fairhaven	260	1863
William Gifford	New Bedford	320	1865
William Hamilton	New Bedford	461	1835
William Lee	Newport	310	1841,54,58,59
Wm. Thompson	New Bedford	495	1839
William Wirt	Fairhaven	386	1840,44,48,49
Young Eagle	Nantucket	377	1841
Zenas Coffin	Nantucket	338	1839,44,45,46,47
Zephyr	New Bedford	361	1853,54
Zone	Nantucket	365	1835

(Source: National Archives, Washington, D.C., Record Group 59, General Records of the Department of State, microfilm, Dispatches from United States Consuls in Paita, 1833-65. Names and tonnages have been checked against Alexander Starbuck's *History of the American Whale Fishery*, as well as W.P.A. transcriptions of customs registrations and various compilations of U.S. merchant vessels. Home ports have been identified from these sources when possible.)

Notes

Introduction

1. Jorge Basadre, *Introduccion a las bases documentales para la historia de la República del Perú; con algunas reflexiones*, 2 vols. (Lima, 1971), 1: 141.

2. Carlos Sempat Assadourian, *El sistema de la economia colonial: mercado interno, regiones y espacio económico* (Lima, 1982), 19 passim.

3. Basadre, *Introduccion a las bases*, 1: 141.

4. Remark made at a panel on "The Frontier Mission in Latin American History" at the twenty-fifth-annual Latin American Studies Association (LASA) meeting in Washington, D.C., 4 April 1991.

5. Paul Gootenberg, *Between Silver and Guano: Commercial Policy and the State in Postindependence Peru* (Princeton, 1989), viii.

Chapter 1

1. Henry Willis Baxley, *What I Saw on the West Coast of South and North America, and at the Hawaiian Islands* (New York, 1865), 61.

2. William H. Prescott, *History of the Conquest of Peru, with a Preliminary View of the Civilization of the Incas*, 2 vols. (New York, 1851), 1: 281-82.

3. Sebastián Lorente, *Historia de la conquista del Perú* (Lima, 1861), 108-10.

4. Prescott, *History of the Conquest*, 1: 358-59; Antonio Raimondi, *El Perú: tomo II, historia de la geografía del Perú* (Lima, 1876), 2: 20; Richard Spruce, *Notes on the Valleys of Piura and Chira, in Northern Peru, and on the Cultivation of Cotton Therein* (London, 1864), 10.

5. On the Chira and Piura River oases, see Claude Collin Delavaud, *Les régions côtiéres du Pérou septentrional; occupation du sol, amenagement regional* (Lima, 1968), 418-54, and Spruce, *Notes on the Valleys*, 6-7, 10, 24.

6. Richard Walters, *A Voyage Round the World, in the Years MDCCXL, I, II, III, IV. by George Anson, Esq; Commander in Chief of a Squadron of His Majesty's Ships, Sent Upon an Expedition to the South Seas...,* (London, 1748), 189; [William S. W. Ruschenberger], *Three Years in the Pacific; Including Notices of Brazil, Chile, Bolivia and Peru* (Philadelphia, 1834), 414; National Archives of the United States, Diplomatic and Consular Records, RG 59, T600, Ringgold to SecState, Paita, 10 January 1854, N 6; Baxley, *What I Saw,* 61; Spruce, *Notes on the Valleys,* 6.

7. Ruschenberger, *Three Years in the Pacific,* 414; Spruce, *Notes on the Valleys,* 7; Baxley, *What I Saw,* 61-62; Berthold Carl Seeman, *Narrative of the Voyage of H.M.S.* Herald *During the Years 1845-51, under the Command of Captain Henry Kellett, R.N., C.B.; Being a Circumnavigation of the Globe,...,* 2 vols. (London, 1853), 146.

8. Ruschenberger, *Three Years in the Pacific,* 414; see the panoramic view of Paita taken from Walters, *A Voyage Round the World,* between 200-01, reproduced in plate 3. On the legend of the wounded Virgin, see William Bennet Stevenson, *A Historical and Descriptive Narrative of Twenty Years' Residence in South America, in Three Volumes* (London, 1825), 2: 196.

9. E. George Squier, *Peru: Incidents of Travel and Exploration in the Land of the Incas* (New York, 1877), 25.

10. Walters, *A Voyage Round the World,* 189; Ruschenberger, *Three Years in the Pacific,* 414; Stevenson, *A Historical and Descriptive Narrative,* 2: 196; Spruce, *Notes on the Valleys,* 12, 19. Walters is the source of the term *baxareque* or *bajareque,* which in Cuba is a small hut or rancho.

11. Walters, *A Voyage Round the World,* 189-90; Baxley, *What I Saw on the West Coast,* 64-65.

12. Spruce, *Notes on the Valleys,* 46; Carlos Sempat Assadourian, *El sistema de la economia colonial: mercado interno, regiones y espacio economico* (Lima, 1982), 294; George Kubler, *The Indian Caste of Peru, 1795-1940. A Population Study Based upon Tax Records and Census Reports* (Washington, D.C., 1952), 34-35; Carlos Deustua Pimentel, *Las intentencias en el Perú (1790-1796)* (Seville, 1965), 117; John Robert Fisher, *Government and Society in Colonial Peru: the Intendent System, 1784-1814* (London, 1970), 253.

13. [International] Bureau of the American Republics, *Peru, Bulletin No 60, 1892* (Washington,D.C., revised to 1 May 1895), 81-82; Kendall W. Brown, *Bourbons and Brandy: Imperial Reform in Eighteenth-Century Arequipa* (Albuquerque, 1986), 99.

14. Spruce, *Notes on the Valleys*, 9. In the late colonial period and most of the nineteenth century the Chira River emptied into the bay of Paita just north of where the town of Colán is located. At some point later in the century, however, the river apparently changed course, creating a new mouth considerably north of Colán. See Map 1 and Tadeo Haenke, *Descripción del Perú* (Lima, 1901), 241.

15. Walters, *A Voyage Round the World*, 189; René-Primevère Lesson, *Voyage autour du monde entrepris par ordre du Gouvernement sur la corvette* La Coquille, originally published in Paris, 1838, reproduced in Peru, Comisión Nacional del Sesquicentenario de la Independencia del Perú, *Colección documental de la independencia del Perú*, tomo XXVII Relaciones de viajeros, 2 vols. (Lima, 1971), 382-33; German Stiglich, *Diccionario geográfico del Perú* (Lima, 1922), 259; Baptismal registry, Parroquia de San Lucas de Colán, in Church of Jesus Christ of Latter Day Saints, Family History Library, microfilm 1507042, items 11-15.

16. After independence, the Province of Piura became part of the Department of La Libertad created in 1837 under the Peru-Bolivian Confederation. President Ramón Castilla's decree of 30 March 1861 separated Piura from La Libertad, and elevated it to the rank of Department, Jorge Basadre, *Historia de la República del Perú*, 5th ed. (Lima, 1963), 3: 1243.

17. Spruce, *Notes on the Valleys*, 10 note; Stevenson, *A Historical and Descriptive Narrative, 198*; Walters, *A Voyage Round the World*, 193; Ruschenberger, *Three Years in the Pacific*, 415.

18. Spruce, *Notes on the Valleys*, 14; for an 1847 description of the road between Paita and Piura and a stopover at Congorá, see Seeman, *Narrative of the Voyage of H.M.S.* Herald, 148-49.

19. Alejandro Garland, *Peru in 1906, with a Brief Historical and Geographical Sketch* (Lima, 1907), 71.

20. Lawrence A. Clayton, *Los astilleros de Guayaquil colonial* (Guayaquil, 1978), 60; Stevenson, *A Historical and Descriptive Narrative*, 2: 198; A. de la Salle, *Voyage autour du monde éxécute pendant les années 1836 et 1837 sur la corvette* La Bonite,..., *relation du voyage*, 3 vols. (Paris, 1851), 2: 63; Seeman, *Narrative of the Voyage of H.M.S.* Herald, 144.

21. From *A Voyage to South America* (Dublin, 1753), quoted in H. J. Mozans, *Along the Andes and Down the Amazon* (New York and London, 1911), 115-16.

22. Walters, *A Voyage Round the World*, 189; Stevenson, *A Historical and Descriptive Narrative*, 113; 177-80.

23. Clayton, *Los astilleros de Guayaquil*, 60; Walters, *A Voyage*

Round the World, 189.

24. Stevenson, *A Historical and Descriptive Narrative*, 75-77.

25. Raimondi, *El Peru*, 278; Stevenson, *A Historical and Descriptive Narrative*, 112.

26. Stevenson, *A Historical and Descriptive Narrative*; William S. Bell, *An Essay on the Peruvian Cotton Industry, 1825-1920* (Liverpool, 1985), 9.

27. Seeman, *Narrative of the Voyage of H.M.S. Herald*, 145.

28. Anson, who had seen early service off the Carolina coast, was made commander of the squadron bound for the Pacific in 1740 with a mission to prey on Spain's colonies. The squadron sailed from St. Helena in September with 960 men. After devastating losses in men and vessels incurred while rounding Cape Horn, the squadron regrouped and rested at Juan Fernández Island, before continuing north to prey on Spanish shipping. After pillaging Paita, the squadron made for China, Walters, *A Voyage Round the World*, and *Dictionary of National Biography*, 2: 31-36.

29. Walters, *A Voyage Round the World*, 187, 194-202.

30. Ibid., 209.

31. John Miller, *Memoirs of General [William] Miller, in the Service of the Republic of Peru*, 2 vols. (London, 1828), 1: 210-11.

32. Gabriel Lafond de Lurcy, *Voyages autour du monde et naufrages célebres,* 2 vols. (Paris, 1843), 2: 152.

33. Walters, *A Voyage Round the World*, 186-87, 177.

34. Ibid., 200-02.

35. *"tan brillante, tan claro como la luna de Paita"* was the refrain, repeated in 1824 by the French traveler, René Lessón, *Voyage autour du monde*, 365.

Chapter 2

1. Richard Walters, *A Voyage Round the World, in the Years MDCCXL, I, II, III, IV, by George Anson, Esq.; Commander in Chief of a Squadron of His Majesty's Ships, Sent Upon an Expedition to the South Seas...*, (London, 1748), 187, 197, 200-01.

2. Robert Proctor, *Narrative of a Journey Across the Cordillera of the Andes, and of a Residence in Lima, and Other Parts of Peru, in the Years 1823 and 1824* (London, 1825), 204-05.

3. David Bushnell and Neill Macaulay, *The Emergence of Latin America in the Nineteenth Century* (New York, 1988), 39.

4. Alexander Starbuck's *History of the American Whale Fishery from its Earliest Inception to the Year 1876*, 2 vols. (1876; reprint, New York, 1964), is the best single source on the history of the New England whalers in the Pacific. See also: Thomas Beale, *The Natural History of the Sperm Whale;...* (London, 1839); Elmo Paul Hohman, *The American Whaleman: Study of Life and Labor in the Whaling Industry* (1928; reprint, Clifton, New Jersey, 1972); Robert Langdon, *American Whalers and Traders in the Pacific: A Guide to Records on Microfilm* (Canberra, 1978); Robert Langdon, *Thar She Went: An Interim Index to Pacific Ports and Islands Visited by American Whalers and Traders in the 19th Century* (Canberra, 1979). The Kendall Whaling Museum in Sharon, Massachusetts, the Old Dartmouth Historical Society Whaling Museum Library in New Bedford, Massachusetts, and Mystic Seaport Museum's G. W. Blunt White Library in Mystic, Connecticut, house rich collections of manuscript material dealing with the whaling industry. Other repositories in New England offer similar resources, and United States consular records of Pacific whaling ports also contain useful material.

5. For a comparison of sperm oil and whale oil prices see Starbuck, *History of the American Whale Fishery,* 660-61. Random comparisons for the years 1810, 1820, 1830, 1840, 1850, and 1860 reveal that the average yearly price for right whale oil for those years was 53, 37, 59, 30, 40 and 35 percent, respectively, of the average yearly price for sperm oil.

6. Zephaniah W. Pease, *History of New Bedford*, 2 vols. (New York, 1918), 1: 366.

7. Starbuck, *History of the American Whale Fishery*, 1: 11-100; Edouard A. Stackpole, *The Sea-Hunters; the New England Whalemen During Two Centuries, 1635-1835* (Philadelphia & New York, 1953), 152-53.

8. Stackpole, *The Sea-Hunters*, 154-55; "The Adventures of John Nicol, Mariner," in *The Sea, The Ship and the Mariner* (Salem, Massachusetts, 1925), 141-45, cited in Edouard A. Stackpole, *Whales & Destiny; the Rivalry between America, France, and Britain for Control of the Southern Fishery, 1785-1825* (Amherst, 1972); National Archives of the United States, Diplomatic and Consular Records, RG 59 (hereafter cited as NA), T600, Ringgold to SecState, Paita, 1 September 1858.

9. *Niles Weekly Register*, 22 July 1815, 368.

10. Ibid., 16 June 1819, 272; 7 August 1819, 400; 14 August 1819, 416.

11. Ibid., 29 January 1820, 382.

12. Ibid., 3 November 1821, 160.

13. Ibid., 10 May 1823, 146; 29 November 1823, 203; 15 February 1823.

14. Starbuck, *History of the American Whale Fishery*, 1: 111.

15. For the impact of the whaling trade on Paita beginning in 1832, see Chapter 3.

16. Zephaniah W. Pease, *History of New Bedford*, 2 vols. (New York, 1918), 2: 302; Daniel Ricketson, *The History of New Bedford, Bristol County, Massachusetts:*... (New Bedford, 1858), 372-73; Russell Bourne, *The View From Front Street: Travels in New England's Historic Fishing Communities* (New York and London, 1989), 175.

17. Starbuck, *History of the American Whale Fishery*, 1: 99-100; NA, T600, Higginson to SecState, Paita, 30 June 1835; *Sailors Magazine and Naval Journal*, 8 June 1836.

18. Starbuck, *History of the American Whale Fishery*, 1: 98.

19. NA, T600, Alexander Ruden, Jr., to SecState, Paita, 14 July 1846, N 55.

20. Starbuck, *History of the American Whale Fishery*, 1: 100. For U.S. sperm oil prices for the period 1832-65, see figure 2.

21. Starbuck, *History of the American Whale Fishery*, 1: 1-2.

22. Journal, ship *Chelsea* of New London, 1831-34, G.W. Blunt White Library, Mystic Seaport Museum (herafter cited as GWBWL).

23. Thomas Boyd, journal, USS *Brandywine*, 1826-28, GWBWL.

24. Thomas Roe, journal, ship *Chelsea* of New London, 1831-34, microfilm 14, GWBWL; William S. W. Ruchenberger, *Three Years in the Pacific; Including Notices of Brazil, Chile, Bolivia and Peru* (Philadelphia, 1834), 415; René-Primevère Lessón, *Voyage autour du monde entrepris par ordre du Gouvernement sur la corvette* La Coquille, originally published in Paris, 1838, reproduced in Comisión Nacional del Susquicentenario de la independencia del Perú, *Colección documental de la independencia del Perú*, tomo XXVII Relaciones de viajeros, 2 vols. (Lima, 1971), 375; A. de la Salle, *Voyage autour du monde éxécute pendant les années 1836 et 1837 sur la corvette* La Bonite...., *Relation du voyage*, 3 vols. (Paris, 1851), 71.

25. Journal, ship *Mercury* of New Bedford, 1837-40, GWBWL.

26. Ruschenberger, *Three Years in the Pacific*, 414-16; de la Salle, *Voyage autour du monde*, 73; Madeleine Vinton Dahlgren, *South Sea Sketches; a Narrative* (Boston, 1881), 233.

27. Lessón, *Voyage autour du monde*, 2: 363.

28. de la Salle, *Voyage autour du monde*, 66; Lesson, *Colección*

documental, 377; Berthold Carl Seeman, *Narrative of the Voyage of H.M.S.* Herald *During the Years 1845-51, under the Command of Captain Henry Kellett, R.M., C.B.; Being a Circumnavigation of the Globe...*, 2 vols. (London, 1853), 146.

29. Tadeo Haenke, *Descripción del Perú* (Lima, 1901), 241; William Bennet Stevenson, *A Historical and Descriptive Narrative of Twenty Years' Residence in South America, in Three Volumes*, (London, 1825), 2: 197; William Funnell, *A Voyage Round the World. Containing an Account of Captain Dampier's Expedition into the South-Seas in the Ship* St. George *in the Years 1703 and 1704* (London, 1707; reprint, Amsterdam, 1969), 175.

30. Lessón, *Voyage autour du monde*, 383; NA, T600, Fayette M. Ringgold to SecState, Paita, 10 January 1854, N 6.

31. Henry Willis Baxley, *What I Saw on the West Coast of South and North America, and at the Hawaiian Islands* (New York, 1865), 64; Alexander George Findlay, *A Directory for the Navigation of the South Pacific Ocean; with Descriptions of its Coasts, Islands, etc. from the Strait of Magalhaens to Panama,...*, 4th ed. (London, 1877), 241; NA, T600, Charles Winslow to SecState, Paita, 9 September 1862, N 2.

32. Lessón, *Voyage autour du monde*, 388; Stevenson, *A Historical and Descriptive Narrative*, 2: 184; NA, 600, Ringgold to SecState, Paita, 10 January 1854, N 6; Findlay, *A Directory for the Navigation*, 241; Richard Spruce, *Notes on the Valleys of Piura and Chira, in Northern Peru, and on the Cultivation of Cotton Therein* (London, 1864), 35, 43-44, 49; Roe, journal, GWBWL; Charles M. Pepper, *Panama to Patagonia: the Isthmian Canal and the West Coast Countries of South America* (Chicago, 1906), 77. Spruce, *Notes on the Valleys*, confirms that prior to the beginning of the cotton boom a significant percentage of the cultivable land in the Chira River valley (the *vegas*) was owned outright and cultivated by Indian and mixed-blood peasants, who were not used to working for wages for hacienda owners, 49, 80.

33. On the virtues of the *algarrobo*, see Cosme Bueno, *Geografía del Perú virreinal (siglo XVIII)* (Lima, 1951), 55; Baxley, *What I Saw on the West Coast*, 64; Findlay, *A Directory for the Navigation*, 241; Spruce, *Notes on the Valleys*, 37. According to numerous accounts and receipts in the manuscript collection of the New Bedford Whaling Museum, whalers calling at Tumbes between 1841 and 1863 paid between $2 and $5 for a *tarea* (cord) of firewood. The tryworks were fired up using firewood and then sustained using the blubber "cracklings" left over from the rendering of sperm oil as fuel. The role of the ubiquitous *algarrobo* in the economic life of the oases of northern Peru

has been neglected. The long pods of the carob tree, which flourished in the river valleys, were a high-calorie, high-protein source of fodder for livestock. As a result, according to Spruce, *Notes on the Valleys*, 37, the *algarrobo* groves were known locally as *potreros* (pastures) rather than *bosques*. The carob bean also was believed to be an aphrodisiac. According to the British botanist, when the first crop of beans ripens and begins to fall from the trees in April, "one cannot walk in the *potreros* without risk of being run over by the amorous donkeys that career [*sic*] madly about."

34. Findlay, *A Directory for the Navigation*, 240.

35. Bueno, *Geografía del Perú*, 51. On the soap industry in colonial Piura see Susana Aldana, *Empresas coloniales: las tinas de jabón en Piura* (Lima, [n.d.]).

36. Stevenson, *A Historical and Descriptive Narrative*, 2: 194; 181-83; Haenke, *Descripción del Perú*, 238.

37. Ruschenberger, *Three Years in the Pacific*, 407.

38. Aldana, *Empresas coloniales*, Appendix 6, 168-70; Stevenson, *A Historical and Descriptive Narrative*, 2: 120.

39. See chapter 5 for a fuller development of the data on American ship calls at Paita taken from consular returns.

40. NA, T600, "Return of the Arrival and Departure of American Vessels, &c." Paita, 31 December 1862; Winslow to SecState, Paita, 31 December 1863, N 16.

41. Findlay, *A Directory for the Navigation*, 241.

42. NA, T600, Ringgold to SecState, Paita, 12 April 1854; Lawrence A. Clayton, *Grace: W.R. Grace & Co., the Formative Years, 1850-1930* (Ottawa, Illinois, 1985), 32.

43. Lawrence A. Clayton, *Los astilleros de Guayaquil colonial* (Guayaquil, 1978), 82.

44. Stevenson, *A Historical and Descriptive Narrative*, 181-83; Aldana, *Empresas coloniales*, Appendix 3, 160-61.

45. Paul Gootenberg, *Between Silver and Guano: Commercial Policy and the State in Postindependence Peru* (Princeton, 1989), 47-48.

46. NA, T600, Ringgold to SecState, Paita, 12 April 1854.

47. [International] Bureau of the American Republics, *Peru, Bulletin, No 60, 1892* (Washington, D.C., revised to 1 May 1895), 97.

48. Clayton, *Los astilleros de Guayaquil*, 81.

49. Ibid., 77-78.

50. Stevenson, *A Historical and Descriptive Narrative*, 2: 194-95; Bueno, *Geografía del Perú*, 55.

51. Bueno, *Geografía del Perú*, 55; Findlay, *A Directory for the*

Navigation, 242; Clayton, *Los astilleros de Guayaquil*, 82, 137; Haenke, *Descripción del Perú*, 240; Spruce, *Notes on the Valleys*, 8.

52. G[abriel] Lafond [de Lurcy], *Voyages autour de monde et naufrages célebres*, 2 vols. (Paris, 1843), 2: 152.

53. Clayton, *Los astilleros de Guayaquil*, 29-30, 79-80; NA T600, Ringgold to SecState, Paita, 12 April 1854.

54. NA, T600, Ringgold to SecState, Paita, 12 April 1854.

55. A daily wage in Paita in the mid-1850s was 75 cents or less for work at the port, and as little as 25 cents in the interior of the Department of Piura, NA, T600, Ringgold to SecState, Paita, 27 November 1854; see also A. de la Salle, *Voyage autour du monde*, 2: 66.

56. Clayton, *Los astilleros de Guayaquil*, 29-30; NA, T600, Ringgold to SecState, Paita, 12 April 1854; Baxley, *What I saw on the West Coast*, 63.

57. Starbuck, *History of the American Whale Fishery*, 1: 112.

58. NA, T600, Ringgold to SecState, Paita, 12 April 1854. On the purchase of *camotes* see MSS collections number 105, 55, 80, 66 at the New Bedford Whaling Museum.

59. For the methodology used to arrive at an approximate number of ship calls by American vessels for the period 1832-65, see chapter 5.

60. Alfonso Rumazo González, *Manuela Sáenz, la libertadora del Libertador* (Bogota, [1945]), 278. I have been unable to find an explanation for the name given to this neighborhood in Paita, beyond the most obvious one that it was frequented by American sailors, or a precedent in United States maritime folklore for the use of "maintop" to identify such a neighborhood.

61. A. de la Salle, *Voyage autour du monde*, 71; Seeman, *Narrative of the Voyage of H.M.S. Herald*, 146.

62. A. de la Salle, *Voyage autour du monde*, 74; James C. Osborn, journal, ship *Charles W. Morgan* of New Bedford, 1841-44, GWBWL.

63. Lessón, *Voyage autour du monde*, 378-79, 390.

64. Journal, ship *Chelsea*, GWBWL; Osborn, journal, GWBWL; Joan Druett, ed., *"She Was A Sister Sailor": The Whaling Journals of Mary Brewster, 1845-51* (Mystic, 1992), 143-51.

65. Stevenson, *A Historical and Descriptive Narrative*, 2: 195.

66. Ibid.

67. See, for example, Bueno, *Geografía del Perú*, 56; Stevenson, *A Historical and Descriptive Narrative*, 2: 195-96; Ruschenberger *Three Years in the Pacific*, 415; and Baxley, *What I Saw on the West Coast*, 65-66; Antonio Raimondi, *El Peru: tomo II, historia de la geografía del Perú*

(Lima, 1876), 278.

68. See, for example, Steven Ruggles, "Fallen Women: the Inmates of the Magdalen Society Asylum of Philadelphia, 1836-1908," *Journal of Social History* 16 (Summer 1983): 65-82; and Susan P. Conner, "Politics, Prostitution and the Pox in Revolutionary Paris, 1789-1799," *Journal of Social History* 22 (Summer 1989): 713-34.

69. See Donna J. Guy, "White Slavery, Public Health, and the Socialist Position on Legalized Prostitution in Argentina, 1913-1936," *Latin American Research Review* 23:3 (1988): 60-80.

70. Ruggles, "Fallen Women," 66.

71. Conner, "Politics, Prostitution and the Pox," 715.

72. David McCreery, in his "'This Life of Misery and Shame': Female Prostitution in Guatemala City, 1880-1920," *Journal of Latin American Studies* 18 (1986): 333-53, reports the following occupations for registered prostitutes in late-nineteenth-century Guatemala City:

Servant/Cook	69
Housewife	48
Seamstress	38
Cigarette/cigar maker	33
Washerwoman	12
Tortilla maker	12

73. This speculation may seem to have racist overtones, particularly when considered in the nineteenth-century context, when many North American and European travelers spoke disparagingly of the results of racial mixture in Latin America. The result, in their view, was the "debasement" of the "white" race, while at the same time they believed that the infusion of European or Anglo-Saxon "blood" into the Latin American gene pool somehow "improved the stock." A candid expression of these values can be found in Baxley, *What I Saw on the West Coast*.

74. Michael C. Meyer and William Sherman, *The Course of Mexican History*, 4th ed. (New York, 1990), 418.

Chapter 3

1. I have ranked the establishment of these consulates by the date of the first despatches sent. See National Archives Trust Fund Board, National Archives and Record Administration, *Diplomatic*

Records; A Select Catalog of National Archives Microfilm Publications (Washington, D.C., 1986).

2. National Archives of the United States, Diplomatic and Consular Records, RG 59 (hereafter cited as NA), T600, Ruden to SecState, Paita, 1 July 1839, N 1; Polhemus to SecState, Paita, 30 June 1848; and Ringgold to SecState, Paita, 1 November 1853.

3. NA, T600, copy, Charles Higginson to John H. Clark, Esq., Paita, 3 June 1839.

4. NA, T600, Ruden to SecState, Paita, 1 January 1841, N 16; 9 February 1846, N 52.

5. NA, T600, Gerard Garland to SecState, Paita, 12 June 1852, No. 85; 29 July 1852, N 86. Peruvian President Echenique had allowed Flores to put together and arm the expedition in Peru, and even contributed personal funds to it, hoping that Flores would unseat the liberal government of José Maria Urbine, which had manumitted Ecuador's slaves, expelled the Jesuits, and instituted freedom of the press. Jorge Basadre, *Historia de la República del Perú*, 5th ed. (Lima, 1963), 2: 970-71; Alfredo Pareja Diezcanseco, *Historia del Ecuador* (Quito, 1962), 274-75.

6. NA, T600, Columbus to SecState, Paita, 18 April 1866, N 7.

7. NA, T600, Columbus to SecState, Paita, 11 January 1868, N 1. On the 1868 revolution and its repercussions in the north see Basadre, *Historia de la República*, 4: 1687, passim. As we shall see in chapter 11, a distant reverberation of Diez Canseco's revolution was to complicate the life of the last American consul in Paita.

8. NA, T600, C.B. Polhemus to SecState, Paita, 15 July 1848, N 62. In August of 1845 the Peruvian government prohibited the export of alpacas, but a few were sent to Australia via England. An attempt at direct shipment to Australia in 1850 failed, Basadre, *Historia de la República*, 2: 847.

9. NA, T600, Garland to SecState, Paita, 12 September 1852, N 88. In the early 1850s, English and American entrepreneurs began to propagate the idea that the Lobos Islands were no-man's-land and that guano shipments from them should be free of Peruvian government regulation. In August of 1852 Secretary of State Daniel Webster advanced the theory that the Lobos were outside the continental waters of Peru and that Lima's assertion of sovereignty was prejudicial to American sealers who had worked those waters for many years. Webster's successor, however, acknowledged Peru's claim to the Lobos and even apologized, Basadre, *Historia de la República*, 2: 968-69.

10. NA, T600, copy of Charles Higginson's declaration, Paita, 27 May 1839; Ruden to SecState, Paita, 14 July 1846, N 55.

11. An 1858 report by the American consul in Paita on the economics of the whaling industry and the real benefits derived by a crew member from a whaling trip reveals that the ordinary sailor profited little from his decision to ship out on a Yankee whaler, see NA, T600, Ringgold to SecState, Paita, 1 January 1858.

12. James Farr, "A Slow Boat to Nowhere: The Multi-Racial Crews of the American Whaling Industry," *Journal of Negro History* 68:2 (1983): 159. For a detailed description of these perils see Alexander Starbuck, *History of the American Whale Fishery from Its Earliest Inception to the Year 1876*, 2 vols. (1876; reprint, New York, 1964), 1: F, "The Dangers of Whale Fishing," 114-45.

13. Thomas Roe, journal, ship *Chelsea* of New London, microfilm, G.W. Blunt White Library, Mystic Seaport Museum.

14. NA, T600, William Rose to SecState, n.p., 28 December 1856.

15. NA, T600, A.A. Atocha to SecState, New York, 11 December 1861; A.A. Atocha to William H. Seward, New York, 10 October 1861; R.W. Griswold to Seward, New York, 10 October 1861.

16. NA, T600, Ringgold to SecState, Paita, 14 April 1860; 16 May 1860; copy, A. G. Blakey to Ringgold, Talcahuano, n.d.

17. NA, T600, Ringgold to SecState, Paita, 10 February 1854, N 7; Charles Harris to SecState, Brooklyn, 17 May 1854. Of all the documented cases of American seamen who "went native," never to return to the United States, that of David Whippey of Nantucket is one of the most intriguing. Whippey jumped ship in Fiji in 1839, became a leader of the community, acquired several wives — in addition to the one left behind in Nantucket — and fathered numerous children. Starbuck, *History of the American Whale Fishery*, 1: 98.

18. NA, T600, C. Mitchell and Company to SecState, Nantucket, 22 September 1849.

19. NA, T600, Ruden to SecState, Paita, 18 June 1842, N 27; 30 June 1842, N 29.

20. NA, T600, Garland to SecState, Paita, 12 October 1851.

21. Sources for these incidents: NA, T600, Garland to SecState, Paita, 24 October 1853; Ringgold to SecState, Paita, 12 December 1853; 15 December 1853; 31 December 1854; Statement of Consular Fees, Paita, 31 December 1854; Ringgold to SecState, Paita, 10 August 1857; 18 July 1857; Samuel J. Oakford to Ringgold, Tumbez, 4 May 1857.

22. Sources for these incidents: NA, T600, Ruden to SecState,

Paita, 10 September 1839; 5 November 1844, N 45; 30 July 1842, N 31; Garland to SecState, Paita, 27 November 1852, N 89.

23. The majority of whalers operating out of New Bedford during its heyday carried several black crewmen, almost always assigned to the more menial, lower-paying jobs. On the role of African-American crew members on whaling ships during this period, see Farr, "A Slow Boat to Nowhere," 164.

24. NA, T600, Ringgold to SecState, Paita, 9 April 1854, NA, T393, Santiago C. Montjoy to SecState, Lambayeque, 13 July 1866.

25. Starbuck, *History of the American Whale Fishery*, 1: 112.

26. Farr, "A Slow Boat to Nowhere," 162; NA, T600, Higginson to SecState, Paita, 18 May 1838; Ruden to SecState, Paita, 1 January 1841, N 13; Ringgold to SecState, Paita, 2 January 1856; 27 November 1854.

27. NA, T600, Winslow to SecState, Paita 6 July 1863.

28. NA, Ringgold to SecState, Paita, 16 June 1860, N 8; 31 December 1854. To solve the problem of stranded seamen, Ringgold worked out an agreement with the local agent of the Pacific Steam Navigation Company that allowed them to travel to Panama or Callao, where they were more likely to find work, for $10. The lowest deck fare to either place was normally $26.50.

29. NA, Winslow to SecState, Paita, 23 January 1863, N 1.

30. Starbuck, *History of the American Whale Fishery*, 1: 112-13.

31. NA, T600, Ruden to SecState, Paita, 2 December 1840, N 11; 10 May 1842, N 25; 16 June 1842, N 26; 28 October 1841, N 20; 29 October 1841, N 21; 30 June 1844, N 41; 30 June 1844, N 42; 13 September 1845, N 49. Most of the death reports do not indicate the cause of death.

32. NA, T600, Ruden to SecState, Paita, 28 October 1841, N 20; 29 October 1841, N 21. I have been unable to identify the exact location of Cabo Blanquillo, but James B. Richardson and Elena B. Decima Zamecnik, "The Economic Impact of Martha's Vineyard Whalers on the Peruvian Port of Paita," *The Dukes County Intelligencer* 10:3 (February 1977): 83, believe it to be at a place now called Cuñuz, about five miles north of Paita, in a semi-circular depression formed by the collapse of the cliff sustaining the *tablazo*.

33. Baxley, *What I Saw on the West Coast*, 62-63.

34. NA, T600, Charles Higginson to SecState, Paita, 30 June 1835.

35. NA, T600, Higginson to SecState, Paita, 30 June 1837.

36. NA, T600, see numerous despatches dating from the pe-

riod 1 July 1840 to 31 March 1854.

37. NA, T600, Ringgold to SecState, Paita, 12 April 1854.

38. Daniel Ricketson, *The History of New Bedford, Bristol County, Massachusetts...* (New Bedford, 1858), 369-70.

39. Consultation with the appropriate section of the National Archives in Washington has convinced me that there is no record group for American consular hospitals in the nineteenth century, even though American hospitals did exist in other consular districts.

40. NA, T600, Ruden to SecState, Paita, 12 September 1852, N 87.

41. See chapter 7.

42. NA, T600, Ringgold to SecState, Paita, 12 April 1854.

43. NA, T600, Winslow to SecState, Paita, 10 August 1863, N 9; Stevenson, *A Historical and Descriptive Narrative*, 2: 104. Guaiacum is the resin of the guayacan tree (family *zygophyllaceae*, genus *officinale* or *sanctum*) which is used to treat gout and rheumatism. On the purchase of sarsparilla and guaiacum by New England whalers, see Kendall Whaling Museum, account book for *Congaree*, May 1847; November 1848; New Bedford Whaling Museum, MSS coll. 105, bark *Mermaid*, August 1856.

44. NA, T600, Alfred Duval to SecState, Paita, 2 December 1861; translation of a petition addressed to Your Excellency, Paita, 16 March 1864 from Manuel León y Seminario; James H. Carsten to SecState, Washington, 26 July 1862.

45. NA, T600, Winslow to SecState, Paita, 9 September 1862, N 2.

46. NA, T600, Winslow to SecState, Paita, 1 October 1862, N 1; 31 December 1862.

47. NA, T600, Consulate of the United States of America at Paita to Domingo Davini, M.D., 31 March 1863.

48. Regarding the Bethlemite hospital in Piura, in January of 1842 the cook of an American whaler was discharged in Paita by his master without funds. When he became ill, the consul sent him to the hospital in Piura, but he deserted the hospital and returned to Paita, where he died on 7 March. He was buried at government expense and his clothing was sold to satisfy his debtors. This unhappy experience may have been one of the catalysts for the founding of the American Hospital in Paita, NA, T600, Ruden to SecState, Paita, 7 March 1842, N 24.

49. Stevenson, *A Historical and Descriptive Narrative*, 1: 40, 219, 255.

50. NA, T600, Alcira de Oakford to SecState, Paita, 16 September 1858; Ruden to SecState, Paita, 9 February 1846, N 52.

51. See chapter 9. This was not an official appointment, however, since the United States did not create a consulate at Lahaina until 1850. Walter Smith, II, *America's Diplomats and Consuls of 1776-1865* (Washington, D.C., 1986), 100.

52. *Dictionary of American Biography*, 2: 60-61.

Chapter 4

1. National Archives of the United States, Diplomatic and Consular Records, RG 59 (hereafter cited as NA), T600, Folger to SecState, Auburn, New York, 1833; 16 April 1833; 4 June 1833.

2. NA, T600, James Guidon Sr. to SecState, Philadelphia, 23 January 1835; 31 January 1835.

3. NA, T600, James Guidon Sr. to SecState, Philadelphia 23 January 1835; 31 January 1835; Charles Higginson to SecState, Paita, 1 January 1835.

4. NA, M639, 6 February 1834 petition included in Representative John Reid to SecState, 28 February 1834.

5. NA, T600, Ruden to SecState, Paita, 1 July 1839, N 1.

6. A footnote to this brief profile of the Rudens as a New York mercantile family appears in Longworth's directory for 1837-38, about five years after Alexander Jr. shipped out for points south. His younger brother Jacques appears then for the first time as an independent merchant. A year later (1838-39) — at the same time Alexander Jr. took over the consulate at Paita — Jacques was doing business as a wine merchant on Pearl Street. The following year his wine shop moved to 157 Laurens Street, while a year later Jacques was purveying "seegars" at 110 Canal Street, just down the block from his father's brokerage house. After Alexander Sr's death in 1846 or 1847, Jacques apparently became the head of the family in New York City. He was listed as head of household in the index to the 1850 census, while a sister Rachel - named no doubt for her grandmother - also figured as head of household. Yet another brother, Emmanuel Ruden, was by this time settled in Buffalo, New York. In *Walker's Buffalo City Directory* for 1842 Emmanuel was listed as boarder at the United States Hotel. Two years later he appeared as a clerk in a local brokerage firm. Emmanuel died at the age of 44 on 28 August 1852, just a year after Alexander Jr. relinquished his post as consul in Paita.

7. On the saturation of the South American market for European goods in the immediate post-independence period, the failure of free-trade liberalism in Peru during this period, and the closing of the Peruvian market to foreign goods, see Paul Gootenberg, *Between Silver and Guano: Commercial Policy and the State in Postindependence Peru* (Princeton, 1989), 13, 18-33, 70; William Lofstrom, *La Presidencia de Sucre en Bolivia* (Caracas, 1987), 366-67. On the collapse of European speculation in mining ventures and the 1825 failure of the London bond market see William Lofstrom, *Dámaso de Uriburu: un empresario minero de principios del siglo XIX en Bolivia* (La Paz, 1982), and Frank Griffith Dawson, *The First Latin American Debt Crisis: the City of London and the 1822-25 Loan Bubble* (New Haven, 1990).

8. NA T600, Ruden to SecState, Valparaiso, 10 June 1838.

9. Gootenberg, *Between Silver and Guano*, 23.

10. NA T600, Higginson to SecState, Paita, 18 February 1839; H. Clark, Captain of the U.S. sloop *Lexington*, to SecState, Guayaquil, 25 June 1839.

11. His brother Emmanuel wrote the State Department in 1841 inquiring about his whereabouts, and adding that they had not heard from him in more than four years, NA T600, Emmanuel Ruden to J.K. Paulding, Buffalo, 2 March 1841.

12. Gootenberg, *Between Silver and Guano*, 9.

Chapter 5

1. Robert Langdon, *American Whalers and Traders in the Pacific: A guide to Records on Microfilm* (Canberra, 1978).

2. National Archives of the United States, Diplomatic and Consular Records, RG 59 (hereafter cited as NA), T600, Winslow to SecState, Paita, 10 August 1863, N 9.

3. The data for 1835 is based on first semester consular returns, adjusted to twelve months, and taking into account the seasonal nature of ship calls and cargo amounts, as abundantly demonstrated in figures for the period 1839 on, when data is more complete.

4. *Encyclopedia Americana*, 1990 ed., 3: 164; 5: 46.

5. The American whalers that called at Paita were not large ships. Annual tonnage data for United States ships calling at Paita for the period 1841-53 show very little fluctuation in ship size, from a low of 254 tons (average for 1853) to a high of 346 (average for 1842).

6. The consular returns for 1840 omit figures for the value of oil cargoes, making it necessary to skip from 1839 to 1841 on figure 1. If, however, we estimate the value of oil cargoes for 1840 on the basis of the size of the cargo and the value for the following year, it would seem likely that the value of oil cargoes in 1840 was over $2 million, as compared with the value for 1857 of $1,756,123.

7. Alexander Starbuck, *History of the American Whale Fishery from its Earliest Inception to the Year 1876, 2 vols.* (1876; reprint, New York, 1964), 1: 112.

8. These returns list previous port of call, destination, the date when the ship was launched, its home port, and the names of the owner and master.

9. A note of caution on the figures for ship calls and value of oil cargoes for 1857: The consular records for the first half of the year are missing. However, from the enumeration of the ship calls listed on the left-hand margin of the report for the final quarter of the year, it is obvious that 68 American ships called at Paita that year, a significant increase from 1853 and 1854 (35 and 37 respectively). My reconstruction of the value of the oil cargo for 1857 may be subject to more skepticism, since I simply doubled the value data available for the last half of the year ($506,274 + $371,787 x 2 = $1,756.123). This figure may be slightly high, but even if it were no higher than the value of the oil cargo for the following year, which is based on more complete returns, the total value for 1857 would probably still represent an all-time record for the entire period 1832-65, excepting 1840, for which no value data is available.

10. NA, T600, Columbus to SecState, Paita, 31 March 1868, N 2; 30 June 1868, N 3; 31 December 1868; Murphy to Second Assistant SecState, Paita, 30 October 1871.

11. Starbuck, *History of the American Whale Fishery,* 1: 100-03.

12. Ibid., 1: 109.

13. The ships papers consulted for this analysis can be located by consulting the respective indices of the two museums cited. They relate to the following vessels: *United States, Constitution, Congaree, Herald, Platina, Champion, Sea Queen, Mermaid, Mary and Susan, Edward, Wave, Aurora, Sacramento, Alfred Gibbs, Hecla,* and *Hector.*

14. The differences between the two ports are worth noting at the outset. Whereas Paita was the principal north Peruvian port of call for American whalers beginning in the 1820s, Tumbes did not become important as a whaling port until the early 1840s. In terms of

the volume of provisions sales to American whalers, Tumbes appears to have eclipsed Paita by the 1850s, although Paita remained an important port of call for supplies and mail. During most of the period 1832-63, whaling vessels preferred to ship water at Tumbes, where crew members could take it directly from the Tumbes River, rather than at Paita, where it had to be purchased. Tumbes also appears to be the port of preference for buying firewood during the 1850s and 1860s. Finally, some items — notably citrus fruit, cacao, broom straw, coconuts, and oysters — were more readily available in Tumbes than in Paita.

15. Elhuyar, who with his brother Juan Jose was the first to isolate tungsten, published this treatise in Madrid in 1825. It is cited and endorsed in Carlos Sempat Assadourian, *El sistema de la economia colonial: mercado interno, regiones y espacio económico* (Lima, 1982), 279-89.

Chapter 6

1. Jorge Basadre, *Historia de la República del Perú*, 5th ed. (Lima, 1963), 2: 513; 3: 1246-47.

2. Ibid., 1: 186.

3. Paul Gootenberg, *Between Silver and Guano: Commercial Policy and the State in Postindependence Peru* (Princeton, 1989) 4-8.

4. W.M. Mathew, *The House of Gibbs and the Peruvian Guano Monopoly* (London, 1981), 15-16.

5. A. de la Salle, *Voyage autour du monde éxecuté pendant les années 1836 et 1837 sur la corvette* La Bonite,...., *Relation du voyage*, 3 vols. (Paris, 1851), 2: 65-66, 75; Gootenberg, *Between Silver and Guano*, 20.

6. National Archives of the United States, Diplomatic and Consular Records, RG 59 (hereafter cited as NA), T600, H Clark to SecState, Guayaquil, 25 June 1839.

7. NA, T600, Polhemus to SecState, Paita, 31 December 1849, N 73; Charles M. Pepper, *Panama to Patagonia: The Isthmian Canal and the West Coast Countries of South America* (Chicago, 1906), 77-78.

8. NA, T600, Polhemus to SecState, Paita, 31 December 1849, N 73; James B. Richardson III and Elena B. Decima Zamecnik, "The Economic Impact of Martha's Vineyard Whalers on the Peruvian Port of Paita," *The Dukes County Intelligencer* 10:3 (February 1977): 69. For some reason, the *Almanache de Gotha*, that indispensable guide to dip-

lomatic and consular representation in the nineteenth century, does not list any consulates in Paita from 1837 on.

9. NA, T600, reference to a draft on the firm Ruden & Co. in Ringgold to SecState, Paita, 17 January 1861; Winslow to SecState, Paita, 8 December 1863, N 15, NA, T393, Santiago C. Montjoy to SecState, Lambayeque, 14 February 1868; NA, T393, *1871 - Nuevo itinerario de los vapores entre Panama, Guayaquil, Payta, Callao, Valparaiso e intermedios,* Callao, 22 March 1871, in NA, T600.

10. William S. Bell, *An Essay on the Peruvian Cotton Industry, 1825-1920* (Liverpool, 1985), 25-6; Richard Spruce, *Notes on the Valleys of Piura and Chira, in Northern Peru, and on the Cultivation of Cotton Therein* (London, 1864), 10. According to Spruce, Arrese y Paredes was heir to the Marqueses de Paredes, not Salinas.

11. Although the contemporary records do not reveal the exact nature of the "engineering studies" for which Duvall was contracted, it is quite probable that they were related to the provision of fresh water for Paita, an area in which he had some expertise. In 1854 Duvall published a *Communication in Relation to a Supply of Water for the City of Baltimore, from the Gunpowder River,* in Baltimore.

12. NA, T600, Alfred Duval to SecState, Paita, 2 December 1861; Winslow to SecState, Paita, 8 December 1863, N 14; Bell, *An Essay on the Peruvian Cotton Industry,* 25.

13. Harry Wright Newman, *Mareen Duvall of Middle Plantation; a Genealogical History of Mareen Duvall,...* (Washington, 1952), 301-30; *Genealogy and Biography of Leading Families of the City of Baltimore and Baltimore County, Maryland...* (New York, 1897), 891.

14. There are four full pages of Duvalls in Robert H. McIntire, *Annapolis Maryland Families* (Baltimore, 1979), 212-16.

15. Newman, *Mareen Duvall of Middle Plantation,* 301-30.

16. Spruce, *Notes on the Valleys,* 25, 45, 63 note, 71, 79.

17. The index to the 1840 census for New York lists several Oakfords living in New York City — the only Oakfords in all of the states of the Atlantic seaboard in 1840 — but attempts to learn more about his family background and his motives for going to Peru have been fruitless.

18. NA, T600, Ringgold to SecState, Paita, 30 March 1858; Columbus to SecState, Paita, 30 November 1866, N 6, NA, T393, Santiago C. Montjoy to SecState, Lambayeque, 27 March 1865.

19. *The Concise Dictionary of National Biography,* (London, 1953), 1: 1230, 639, 91; NA, T600, copy, Richard Spruce to William Alexander Cox Cole, Piura, 22 March 1863; Winslow to SecState, Paita,

30 June 1864 N 14.

20. NA, T600, Winslow to SecState, Paita, 30 June 1864, N 14. On Urbina's attempts to overthrow Garcia Moreno see Alfredo Pareja Diezcanseco, *Historia del Ecuador* (Quito, 1962), 275-91.

21. On Manuela's exile in Paita and the visit of Rodríquez see Alfonso Rumanzo Gonzalez, *Manuela Saenz, la libertadora del Libertador* (Bogota, [1945]), 265-86.

22. Spruce, *Notes on the Valleys*, 50; NA, T600, Ringgold to SecState, Paita, 31 December 1853, N 5; Winslow to SecState, Paita, 24 April 1863, N 3; Columbus to SecState, Paita, 15 February 1864; 30 June 1868, N 3; Baxley, *What I Saw on the West Coast of South and North America, and at the Hawaiian Islands* (New York, 1864), 65; John Reid to SecState, Paita, 30 March 1860; Winslow to SecState, Paita, 31 December 1862, N 8; Winslow to SecState, Washington, 23 January 1864; NA, Record Group 84, "Record of Relief Afforded Destitute American Seamen at the United States Consulate, Paita, Peru," January 1857-1874; New Bedford Whaling Museum, MSS coll. 105, bark *Aurora*, bark *Platina*; MSS coll. 55, bark *Mary and Susan*.

23. NA, T600, copy, John Brown to C.F. Winslow, Paita, 23 April 1863; Winslow to SecState, Paita, 24 April 1863, N 3. In maritime law, barratry is a fraudulent breach of duty or willful act of known illegality on the part of the master or crew of a ship to the injury of the owner or cargo.

24. E. Nichols to Mssrs. Wood & Nye, Paita, 1 October 1860 in New Bedford Whaling Museum, MSS coll. 66, bark *Alfred Gibbs*.

25. William S. W. Ruschenberger, *Three Years in the Pacific; Including Notices of Brazil, Chile, Bolivia and Peru* (Philadelphia, 1834), 405.

26. NA, M650, Montjoy to William H. Seward, Lambayeque, 23 March 1866; NA, T393, Montjoy to SecState, Lambayeque, 15 September 1866, N 9; 14 February 1868; Montjoy to Second Assistant SecState, Lambayeque, 28 September 1871; William F. Fry to Second Assistant SecState, Lambayeque, 6 February 1872. The Cokes and the Mountjoys were both prominent Virginia families in the eighteenth and nineteenth centuries, but little is known about their Vermont contemporaries. Published census indices for 1810 through 1840 for Vermont list no Cokes or Montjoys (or Mountjoys). The 1820 index for New York lists a Kimble Coke resident in Schoharie County. There are also several Cokes in the 1840 index, including two in New York City. New York University's medical school functioned on Broadway and graduated its first class of doctors in 1842, but the mysterious Dr.

Montjoy is not listed in the university archives. Nor did he study at Columbia University's College of Physicians and Surgeons.

27. Essex Institute, *Vital Records of Roxbury Massachusetts to the End of the Year 1849*, 2 vols. (Salem, 1925), 1: 109; Francis S. Drake, *The Town of Roxbury; Its Memorable Persons and Places; Its History and Antiquities...* (Boston, n.d.), 27, 29, 60, 82, 283, 286, 343; Walter Eliot Thwing, *History of the First Church in Roxbury, Massachusetts, 1630-1904* (Boston, 1908), 190; Robert P. Spindler, "Synopsis of the Dorr Family Papers in the Phillips Library, Peabody Essex Museum, Salem, Massachusetts," May 1988; *A Report of the Record Commissioners of the City of Boston, Containing Boston Births from A.D. 1700 to A.D. 1800* (Boston, 1894), 303. Ebenezer and Abigail Dorr had ten children, the oldest of which was Ebenezer Jr., Frederick's father, and the youngest of which was Sullivan (1778-1858). Ebenezer Jr., who was born in Boston, apparently moved back to the family estate in Roxbury, where his children were born. Unlike his younger brothers Joseph, John, Andrew, and Sullivan, he was not involved directly in his father's shipping business, although he did own part interest in one of the Dorr ships that traded with Canton.

28. Spindler, "Dorr Family Papers"; Howard Corning, "Sullivan Dorr, China Trader," *Rhode Island History* 3:3 (July 1944): 81; Louis Clinton Nolan, "The Relations of the United States and Peru with Respect to Claims, 1822-1870," *Hispanic American Historical Review* 17:1 (February 1937): 36. After returning from Canton, Sullivan Dorr settled in Providence, Rhode Island, where he became president of the Washington Insurance Company and served as a trustee of Brown University. Sullivan's claims against the Peruvian government for losses associated with the seizure of the *Esther* in 1823 were made possible by a claims convention concluded by the governments of Peru and the United States in 1841, ratified after many delays on both sides in 1845 and "proclaimed" in 1847, Nolan, "Relations of the United States and Peru," 33-34.

29. Regrettably, Dorr family papers in the Rhode Island Historical Society, the Massachusetts Historical Society, and the Burton Collection of the Detroit Public Library contain no correspondence relating to the activities of Sullivan and Frederick Dorr in Peru in the 1830s and 1840s.

30. NA T600, copy of Charles Higginson's declaration, Paita, 27 May 1839; German Stiglich, *Diccionario geográfico del Perú* (Lima, 1922), 1141.

31. George Adams, *The Roxbury Directory: 1852* (Roxbury,

1852); letter, Philip Bergen, The Bostonian Society, Boston, 28 January 1991. Curiously, the "rubber factory" on Cabot Place in Roxbury appears to have been located on land that once formed part of Frederick's grandfather's estate, which "extended from Eustis Street to a point opposite Vernon" Drake, *Town of Roxbury*, 149; *A Map of the City of Roxbury*, from Surveys by C.H. Poole, revised for 1858, in *Roxbury Directory for 1858*, courtesy of Philip Bergen. Philip and Frederick Dorr lived in separate residences only three or four blocks from the rubber factory.

 32. Bell, *An Essay on the Peruvian Cotton Industry*, 23.

 33. Letter from the Bishop of Piura to Ricardo Herrera, Curate of Paita, Piura, 24 June 1941, in Church of Jesus Christ of Latter Day Saints Family History Library (hereafter cited as LDS), microfilm 1507042.

 34. LDS, microfilm 1507042, items 1 through 10.

Chapter 7

 1. (Mrs.) D.B. Bates, *Four Years on the Pacific Coast* (Boston, 1860), 73-79, quoted in James B. Richardson III, and Elena B. Decima Zamecnik, "The Economic Impact of Martha's Vineyard Whalers on the Peruvian Port of Paita," *The Dukes County Intelligencer* 10:3 (February 1977): 74-75.

 2. National Archives of the United States, Diplomatic and Consular Records, RG 59 (hereafter cited as NA), T600, Ruden to SecState, 1 January 1848, N 61; Polhemus to SecState, Paita, 30 June 1848, N 63; Ruden to SecState, Lima, 13 December 1848, with copy of Ruden to SecState, Paita, 13 November 1848; 12 February 1851; Walter B. Smith II, *America's Diplomats and Consuls of 1776-1865*, U.S. Department of State, Foreign Service Institute, Center for the Study of Foreign Affairs, Occasional Paper No. 2 (Washington, D.C., 1986), 122.

 3. NA, T600, "Consular Return of American Vessels...," Paita, 30 June 1852.

 4. The most complete information on the Ringgold family, and the source of the unattributed data on Ringgold and his brothers, is found in Mildred Cook Schoch, *Ringgold in the United States* (privately printed, 1970).

 5. Thomas J.C. Williams, *A History of Washington County, Maryland, from the Earliest Settlements to the Present Time...*, 2 vols. (n.p., 1906); *Washington County, Maryland Church Records of the 18th Century*

(Westminster, Maryland, 1988); Talbot Hamlin, *Benjamin Henry Latrobe* (New York, 1955); several authors, *First Two Centuries of the Washington County Courthouse*. The manor house, Fountain Rock, is said to have been designed by Benjamin Latrobe (1764-1820), architect of the rebuilding of the U.S. Capitol and of St. John's Church on Lafayette Square, but there is no documentation to confirm the Latrobe connection to the house, which burned in 1826. Nevertheless, while in Washington in 1817, Ringgold was instrumental in commissioning Latrobe to prepare the design for the first Washington County Court House (1822-71). Samuel's brother Tench Ringgold, who was one of the three commissioners charged with supervising the rebuilding of the Capitol after it was burned by the British in August 1814, and who worked closely with Latrobe in the reconstruction project, probably introduced Latrobe to his brother during the latter's periodic residences in Washington as a member of Congress.

6. Homer A. Walker, *Historical Court Records of Washington District of Columbia*, vols. 8-11, marriages Joy, T. - Rophinot (n.p., n.d). Unfortunately, District of Columbia marriage records for the period 1811-70 contain only the names of the spouses and the date of marriage.

7. Other accounts identify the bride as the daughter of President James Madison, who of course died without issue. The same accounts suggest that they were married at the White House during the tenancy of President Madison, but all of these stories are probably apocryphal. President Monroe's younger daughter Elizabeth Gouvernor Monroe was married to a widowed Virginia lawyer named George Hay, but their only daughter was named Hortense and the time period does not appear to coincide. Hay did not have any daughters by his first wife, "Hay, George (1765-1830)," in *Dictionary of American Biography*, 8: 430.

8. Were it not for the fact that the Virginia census of 1800 was burned, it might be possible to determine from which of the dozen or so Virginia branches of the Hay family Marie Antoinette was descended.

9. Williams, *A History of Washington County*, 198. A letter written by Mrs. Ringgold in 1853 shows that she was well versed in the art of letter writing and that she had a clear, "educated" hand, NA, M967, M[arie] A[ntoinette] Tidball to Thomas Ritchie, Washington, 17 August 1853.

10. One of Fayette's sisters, Virginia, married John Ross Key, the son of Francis Scott Key — of "Star Spangled Banner" fame —

and Mary Tayloe Lloyd Key, descended from the wealthy and influential Tayloes of Virginia and the equally influential Maryland banking and landowning dynasty of the Lloyds. John Ross Key died in 1837 at the age of 28, just two months before the birth of his third son, John Ross Key, Jr. His widow, Virginia Ringgold Key, lived until 1902, Mrs. Julian C. [Janie W.] Lane, *Key and Allied Families* (Macon, Georgia, 1931), 61.

11. The State of Maryland did not require that birth records be kept until late in the nineteenth century, forcing genealogists to rely on church baptismal records. Unfortunately, the records for St. John parish in Hagerstown were destroyed by fire.

12. Samuel's remains were interred in the family cemetery at Fountain Rock. In 1842 the manor house and some 32 acres were sold to the Episcopal Diocese of Maryland for $5,000 to house the recently created College of St. James.

13. The census does not seem to account for Marie Antoinette's third daughter, who may have been visiting relatives. All three lived to adulthood and married.

14. The only clue that I have been able to find as to the possible existence of Fayette's two younger half brothers is in *Boyd's Washington and Georgetown Directory* for 1858, 302, where an Edward M. Tidball and a J.C. Tidball are listed. Edward worked at the Navy Department and boarded at a house on 15th Street, N.W., while J.C. worked for the U.S. Coast Survey, and boarded on "K" Street. The two boys listed in the 1840 Maryland census as under five would, in 1858, have been in their early twenties. The census data suggests that Marie Antoinette had two male children by Tidball, since there were two boys under five living with them in the 1840 count. In addition to Marie Antoinette, who was 50 at the time, there were three other females in the household between the ages of five and 30, and an elderly free black female.

15. *The Washington Directory, and Governmental Register, for 1843;...*, comp. Anthony Reintzel (Washington, D.C., 1843); *The Washington Directory, and National Register, for 1846...* (Washington, 1846), 80. There were no directories published for Washington for several years prior to 1843, or between the years 1843 and 1846. The directory for 1850 lists Mrs. Tidball's oldest son, Major George H. Ringgold, living on "F" Street, N.W., between 19th and 20th, just a block from his uncle Tench Ringgold's house at "F" and 18th, N.W., but Mrs. Tidball is no longer listed independently. Since she lived the later years with her son George, it is probable that by 1850 she had been incor-

porated into his household, *The Washington Directory, and Congressional and Executive Register, for 1850,* comp. Edward Waite (Washington, D.C., 1850), 74.

16. Williams, *A History of Washington County,* 112.

17. *The Washington and Georgetown Directory, Strangers' Guide Book for Washington,...for 1853,* 57, listed a "Mrs. John Key, widow" living on "H" Street NW, between 15th and 16th.

18. Tench Ringgold was a wealthy and prominent Washingtonian who owned a tannery on the banks of Rock Creek near Georgetown. In addition to his role in rebuilding the Capitol and the White House, Tench Ringgold was Marshall of the Circuit Court of the District of Columbia and Treasurer of the Georgetown Savings Institution. Ringgold owned large tracts of land in the Foggy Bottom section of Washington. In 1825 he built an impressive Federal-style house at the corner of "F" and 18th Streets, N.W., just two blocks from the White House, on land that he had purchased the previous year from the widow of President Washington's private secretary. United States Chief Justice John Marshall lived for a time in this house, which was enlarged and modified in the late nineteenth century and today is known as the DACOR Bacon House.

19. Maximilian Schele De Vere, *Students of the University of Virginia* (Baltimore, 1878), 98; *University of Virginia Catalogue,* 1846, 6; and University of Virginia Matriculation Records for 1845, cited in a letter from Kimberly M. Johnson, 8 February 1991.

20. *Catalogue of the Trustees, Officers, & Students of the University of Pennsylvania, Session 1846-47* (Philadelphia, 1847), 23; letter from Gail M. Pietrzyk, University of Pennsylvania Archives and Records Center, 25 January 1991.

21. NA, M967, J.M. Mason to SecState, Richmond, 15 August 1853; NA T328, Ringgold to SecState, Arica, 5 January 1851; 5 December 1849.

22. As we have seen, both Samuel and Tench Ringgold were Jeffersonian Democrats and were very well connected in the political milieu of early-nineteenth-century Washington. Fayette's mother was also well connected. In a letter to Thomas Ritchie, the founding editor of the Richmond *Enquirer* and an ardent Jeffersonian Democrat, in which she pleads her son's candidacy for the consulate in Paita, she refers to a conversation with President Franklin Pierce about her son's interest in the position. Ritchie, at the request of President Polk, had been editor of the administration organ *Union* from 1845 to 1851, and was still living in Washington when Mrs. Tidball wrote to him on

her son's behalf, NA, M967, M.A. Tidball to Thomas Ritchie, Washington, 17 August 1853. On Ritchie, see *Dictionary of American Biography* (New York, 1935), 15: 628-29.

23. Major Samuel Ringgold, who was educated at West Point, was mortally wounded at the Battle of Palo Alto in the Mexican War and died in Texas in 1846, the year before Fayette graduated from the University of Pennsylvania. Samuel's death made him a hero and martyr; Currier & Ives eulogized him with a gruesome lithograph of his fatal wounding, and towns in Georgia, Louisiana, Pennsylvania, Texas, and Virginia were named for him. Colonel George Hay Ringgold, the oldest of Samuel's children by his second wife, was also a West Point graduate. A gifted scholar and amateur poet, Ringgold served in both the Mexican and Civil Wars and died in 1864 in San Francisco, where his mother was living with him and his family. Rear Admiral Cadwalader Ringgold cruised in the *Vandalia* to the Pacific in 1828-32, and in the *Porpoise* in 1838-42 as part of Wilkes's U.S. Exploring Expedition, which surveyed the coast of Alta California. In 1849 and 1850, at the same time that Fayette signed on as medical officer of the navy storeship, Cadwalader engaged in further surveys of the California coast, a cruise that may have taken him, en route, to Paita. In 1853-54 he served as United States Special Diplomatic Agent in Canton, China, Walter B. Smith II, *America's Diplomats and Consuls of 1776-1865,* 212.

24. NA, M967, J.M. Mason to SecState, Richmond, 15 August 1853; NA, T600, Winslow to SecState, Paita, 24 April 1863, N 3; M.A. Tidball to Thomas Ritchie, Washington, 17 August 1853; NA, T328, Ringgold to SecState, Arica, 15 May 1852; NA T967; Paita, 14 October 1852; NA, T600, Ringgold to SecState, Paita, 29 October 1853; NA, T600, Ringgold to Marcy, Paita, 29 October 1853; United States Treasury receipt, February 1854.

25. NA, T600, Ringgold to SecState, Paita, 1 November 1853.

26. NA, T600, Ringgold to SecState, Paita, 12 April 1854.

27. NA, M967, Ringgold to President Pierce, Paita, 15 January 1857. Pierce was cast aside by his party for renomination in 1856 because he was not anti-slavery enough.

28. Richard Spruce, *Notes on the Valleys of Piura and Chira, in Northern Peru, and on the Cultivation of Cotton Therein* (London, 1864), 78.

29. Smith, *America's Diplomats and Consuls,* 12.

30. NA, T600, Winslow to Abraham Lincoln, Paita, 6 July 1863; Ruden to SecState, Paita, 1 January 1840, N 3; and 23 subse-

quent despatches through 31 December 1853. The consul's privilege of duty-free import of foreign merchandise was not a particular advantage in Paita, which was a wide-open port with only minimal official Peruvian control. In addition, every whaler that called at Paita in the 1830s and perhaps later was allowed to trade, duty free, imported goods with a value of $200. This limitation was often honored in the breach, William S. W. Ruschenberger, *Three Years in the Pacific; Including Notices of Brazil, Chile, Bolivia and Peru* (Philadelphia, 1834), 416.

31. NA, T600, Winslow to SecState, 24 April 1863, N 3.

32. NA, T600, Winslow to SecState, Paita, 1 October 1862, N 1; Ringgold to SecState, Paita, 1 January 1856. Although Lañas is a well-known surname in the Piura region even today, the only notice I have been able to find of Ringgold's father-in-law was this reference to a draft drawn by Ringgold on "Jose Lañas, Esq. of Paita" for the settlement of an account.

33. NA, T600, Ringgold to SecState, Paita, 2 July 1861.

Chapter 8

1. Paul Gootenberg, *Between Silver and Guano: Commercial Policy and the State in Postindependence Peru* (Princeton, 1989), viii.

2. William S. Bell, *An Essay on the Peruvian Cotton Industry, 1825-1920* (Liverpool, 1985), 8.

3. National Archives of the United States, Diplomatic and Consular Records, RG 59 (hereafter cited as NA), T600, Ringgold to SecState, Paita, 12 April 1854.

4. Carlos Sempat Assadourian, *El sistema de la economica colonial: mercado interno, regiones y espacio económico* (Lima, 1982), 223.

5. Assadourian, *El sistema de la economica*, 233.

6. Gootenberg, *Between Silver and Guano*, 19-20.

7. NA, T600, Ringgold to SecState, 12 April 1854.

8. NA, T600, Ringgold to SecState, 12 March 1854.

9. NA, T600, Columbus to SecState, Paita, 24 September 1870, N 7.

10. NA, T600, "Return of the Arrival and Departure of American Vessels, &c.," Paita, 31 December 1862; "Arrivals and Departures of American Vessels at the United States Consulate at Paita," 31 December 1860; "Navigation and Commerce of the United States at the Port of Paita," 30 June 1860; 31 March 1860, 31 March 1859; "Quar-

terly Return of the Arrival and Departure,..." 30 September 1858; Bell, *An Essay on the Peruvian Cotton Industry*, 24.

11. NA, T393, Montjoy to SecState, Lambayeque, 30 June 1865; 31 June 1868, N 32, 30 September 1872.

12. W. M. Mathew, *The House of Gibbs and the Peruvian Guano Monopoly* (London, 1981), 10-11, 19; Gootenberg, *Between Silver and Guano*, 13-14.

13. NA, T600, Winslow to SecState, Paita, 8 December 1863, N 15.

14. NA, T600, Columbus to SecState, Paita, 30 September 1868, N 4; 31 December 1868, N 9; Murphy to Second Assistant SecState, Paita, 30 October 1871.

15. By way of comparison, the famous straw hats of Xipijapa, in the province of Guayaquil, sold for upwards of $20 each during the very late colonial period, William Bennet Stevenson, *A Historical and Descriptive Narrative of Twenty Years' Residence in South America, in Three Volumes* (London, 1825), 2: 234.

16. NA, T600, Ringgold to SecState, Paita, 10 January 1854, N 6; 27 November 1854; Columbus to SecState, Paita, 30 September 1868, N 4; Murphy to Second Assistant SecState, Paita, 30 October 1871. Claude Collin Delavaud, *Les régions côtiéres du Pérou septentrional; occupation du sol, amenagement regional* (Lima, 1968), 293, describes Catacaos as the largest, most homogenous and most combative indigenous community of all of Peru's northern coast.

17. Tadeo Haenke, *Descripción del Perú* (Lima, 1901), 240; Stevenson, *A Historical and Descriptive Narrative*, 2: 167; William S. W. Ruschenberger, *Three Years in the Pacific; Including Notices of Brazil, Chile, Bolivia and Peru* (Philadelphia, 1834), 415; NA, T600, Ringgold to SecState, Paita, 27 November 1954; Columbus to SecState, Paita, 30 September 1868, N 4; 24 September 1870, N 7. On the attempted revival of regional economic integration with Chile, see Gootenberg, *Between Silver and Guano*, 38-46.

18. Henry Willis Baxley, *What I Saw on the West Coast of South and North America, and at the Hawaiian Islands* (New York, 1865), 63-64; Ruschenberger, *Three Years in the Pacific*, 415; NA, T600, Ringgold to SecState, Paita, 12 April 1854; Columbus to SecState, Paita, 30 September 1868, N 4; Murphy to Second Assistant SecState, Paita 30 October 1871.

19. NA, T600, Winslow to SecState, Paita, 8 December 1863, N 15; NA, T393, Montjoy to Second Assistant SecState, Lambayeque, 30 September 1872.

204

20. NA, T600, Ringgold to SecState, Paita, 27 November 1854.

21. On cotton cultivation in pre-conquest Peru, see Collin Delavaud, *Les régions côtiéres du Pérou*, 276-81, and Bell, *An Essay on the Peruvian Cotton Industry*.

22. Richard Spruce, *Notes on the Valleys of Piura and Chira, in Northern Peru, and on the Cultivation of Cotton Therein* (London, 1864), 47.

23. Stevenson, *A Historical and Descriptive Narrative*, 2: 107-08, 183.

24. NA, T600, Winslow to SecState, Paita, 8 December 1863, N 15; Spruce, *Notes on the Valleys of Piura and Chira*, 52, 56.

25. NA, T600, Winslow to SecState, Paita, 8 December 1863, N 15.

26. NA, T600, Ringgold to SecState, Paita, 10 January 1854, N 6; Bell, *An Essay on the Peruvian Cotton Industry*, 8-9. The pre-conquest inhabitants of Peru considered brown cotton to be sacred, and used batting made from the brown fibers to encase their mummies before wrapping them in cotton cloth, Spruce, *Notes on the Valleys of Piura and Chira*, 47.

27. Spruce, *Notes on the Valleys of Piura and Chira*, 50; Baxley, *What I Saw on the West Coast*, 63-64.

28. Spruce, *Notes on the Valleys of Piura and Chira*, 52; NA, T600 Winslow to SecState, Paita, 8 December 1863, N 15.

29. Spruce, *Notes on the Valleys of Piura and Chira*, 55-56.

30. Ibid., 78.

31. NA, T600, Ringgold to SecState, Paita, 12 April 1854; Winslow to SecState, Paita, 8 December 1863, N 15.

32. NA, T600, Murphy to Second Assistant SecState, Paita, 30 October 1871; [International] Bureau of the American Republics, *Peru, Bulletin No. 60*, 1892 (Washington, D.C., revised to 1 May 1895), 49-50; Bell, *An Essay on the Peruvian Cotton Industry*, 29-31; NA T393, Montjoy to SecState, Lambayeque, 30 September 1867.

33. NA, T600, Murphy to Second Assistant SecState, Paita, 30 October 1871.

34. Alejandro Garland, *Peru in 1906, with a Brief Historical and Geographical Sketch* (Lima, 1907), 182; Bell, *An Essay on the Peruvian Cotton Industry*, 31 passim.

Chapter 9

1. National Archives of the United States, Diplomatic and Consular Records, RG 59 (hereafter cited as NA), T600, Ringgold to SecState, Paita, 15 April 1859; 1 January 1860, N 1; John Reid to SecState, Paita, 30 March 1860.

2. NA, T600, Alfred Duval to SecState, Paita, 2 December 1861; Ringgold to SecState, Panama, 23 November 1861; Paita, 30 January 1862.

3. NA, T600, Winslow to SecState, 24 April 163, N 3.

4. NA, T600, Winslow to SecState, Paita, 9 September 1862, N 2 [*sic*]; 1 October 1862, N 1.

5. NA, T600, Winslow to SecState, Paita, 9 September 1862, N 2.

6. NA, T600, Winslow to SecState, Paita, 24 April 1863, N 3; 6 July 1863.

7. Unless otherwise noted, biographical information on Winslow is taken from David-Parsons Holton, *Winslow Memorial. Family Records of Winslows and their Descendants in America...* (New York, 1888), 810-14.

8. Joan Druett, ed., *"She Was a Sister Sailor": The Whaling Journals of Mary Brewster, 1845-51* (Mystic, Connecticut, 1992), 206-08.

9. Charles Winslow, "Letters from California," *The Magazine of History* 103 (1924).

10. See *The Preparation of the Earth for the Intellectual Races: A Lecture Delivered at Sacramento, California, April 10, 1854 at the Invitation of the House of Assembly* (Boston, 1854); and *The Nazarite's Vow. An Address delivered Before the Sons of Temperance, in San Francisco, 4 March 1855* (Boston, 1855).

11. Holton, *Winslow Memorial*, 2: 812.

12. NA, T600, Winslow to Abraham Lincoln, Paita, 14 November 1862; Holton, *Winslow Memorial*, 2: 806; *Encyclopaedia Brittanica*, 6: 67; Winslow's scientific bent could have been inherited from a maternal ancestor, the English mathematician, astronomer, and clergyman Jeremiah Horrocks (ca. 1617-41), who applied Kepler's laws of motion to the Moon and whose observations of a transit of Venus in 1639 are the first recorded. In showing the Moon's orbit to be approximately elliptical, Horrocks laid a partial groundwork for Sir Isaac Newton's later work.

13. NA, M967, Elisha H. Allen to Charles F. Winslow, Washington, 16 December 1856; Winslow to President Buchanan, Wash-

ington, 23 April 1857; John E. Woll to SecState, Washington, 23 April 1857; Garrison and Charles A. Whitney to SecState, Washington, 23 April 1857. Garrison was a New York merchant who went to Panama at the time of the California gold rush, where he opened a thriving commercial and banking house. From there he became a very successful businessman and civil leader in San Francisco, prior to returning to the world of finance in New York, *Dictionary of American Biography*, 7: 167-68.

14. NA, T600, Charles Ingersoll to SecState, Washington, 18 January 1862; Governor Andrew was responsible for the organization of the famous 54th Massachusetts Infantry Regiment of black soldiers, whose heroic exploits were recently the subject of the Oscar-winning film *Glory*.

15. *Dictionary of American Biography*, 1: 279; 9: 465.

16. The trade in Polynesians occurred during the interval between the first importations of Chinese coolies to Peru (1849-54) and the second phase of that traffic (1861-74). On the traffic in Polynesians, see Watt Stewart, *Chinese Bondage in Peru: A History of the Chinese Coolies in Peru, 1849-1874* (Durham, North Carolina, 1951), 28-29; Jorge Basadre, *Introduccion a las bases documentales para a historia de la República del Perú; con algunas reflexiones*, 5th ed. (Lima, 1971), 3: 1450-51.

17. NA, T600, Winslow to SecState, Paita, 31 December 1862, N 9.

18. NA, T600, G.H. Greene to D.H. Greene, Paita, 28 October 1862; G.H. Greene to Allen Greene, Paita, 12 January 1863; George Woodruff to SecState, Washington, 23 February 1863.

19. NA, T600, "El Consulado Americano en Paita," with forwarding note dated Paita, 24 April 1863; John H. Morehouse to SecState, San Francisco, 19 March 1863. Ship papers from the *Alfred Gibbs* and the *Mary and Susan*, in MSS colls. 55 and 66 of the New Bedford Whaling Museum, confirm that in 1863 Dr. Winslow did sell onions, shoes, and sarsaparilla root to American whalers, in addition to providing medical treatment for their crews.

20. NA, T600, Winslow to SecState, Paita 24 April 1863; Certified copies, Richard Spruce to William Alexander Cox Cole, Piura, 22 March 1863; John Brown to C.F. Winslow, Paita, 23 April 1863; Marco Antonio Grisolle to C.F. Winslow, Paita, 24 April 1863.

21. NA, T600, William Wood to SecState, Paita, 29 September 1863.

22. NA, T600, Winslow to SecState, Paita, 15 November 1863.

23. NA, T600, Winslow to SecState, Paita, 7 December 1863, N 14.

24. NA, T600, Winslow to SecState, Paita, 7 December 1863, N 14; 1 January 1864, N 18; Columbus to SecState, Paita, 1 January 1864; 24 October 1865; Winslow to SecState, Boston, 25 January 1864; New Bedford Whaling Museum, MSS coll. 55, barks *Hecla* and *Mary and Susan*.

25. NA, T600, Governor John Andrew to SecState, Boston, 1 February 1864; Winslow to SecState, Washington, 21 January 1864. Blacker's affadavit is dated Paita, 10 December 1863.

26. NA, T600, Winslow to SecState, Boston, 2 February 1864; Paita, 1 March 1864, N 3; 16 March 1864, N 4; 1 April 1864, N 10; 30 June 1864, N 14.

27. NA, T600, Winslow to SecState, Paita, 16 July 1864, N 15; 16 July 1864, N 16; J.M. Hood to J.F. Farnsworth, Syracuse, 2 May 1864; Columbus to SecState, Paita, 1 August 1864, N 1.

28. NA, T600, Wetmore to SecState, Cuyohoga Falls, Ohio, 5 September 1864; 12 September 1864; 10 September 1864; Paita, 15 November 1864; 30 November 1864, N 1; 29 November 1864; 16 January 1865; 2 March 1865; Haven to SecState, Paita, 10 April 1865, N 4.

29. NA, T600, Columbus to SecState, Paita, 24 October 1865; copy of Columbus's naturalization certificate dated 8 April 1865; 28 August 1865 bond signed by John E. Russell, Leonard S. Sanford, and George K. Otis; NA, M650, "letters of Application and Recommendation During the Administrations of Abraham Lincoln and Andrew Johnson, 1861-1869," cover note signed by George Kohs, New York, 22 April 1865.

30. NA, T600, Winslow to SecState, Boston, 19 October 1864; 4 January 1865; 16 February 1865.

31. NA, T600, Quarterly Report dated 30 September 1863; Murphy to Second Assistant SecState, Paita, 31 December 1872, N 18, enclosure.

32. NA, T393, Elisha L. Mix, Jr., to SecState, Lambayeque, 6 January 1864; Montjoy to SecState, Lambayeque, 31 March 1867.

33. NA, T600, Winslow to SecState, Paita, 1 October 1863, N 10; voucher, Treasury Department, Fifth Auditor's Office, Consulate at Paita, 28 December 1863.

34. NA, T600, Winslow to SecState, Paita, 30 June 1863; 31 March, 1864; Wetmore to SecState, Paita, 31 December 1864, N 2.

35. NA, T600, Columbus, "late United States Consul," to Second Assistant SecState, Paita, 23 June 1871.

36. NA, T600, Columbus to SecState, Paita, 30 September 1864, N 3; 16 March 1867, N 9; 31 March 1868, N 2. One variant of cholera, known as bilious cholera, is characterized by bile in the feces. Although Paita may have been spared from the 1868 yellow fever epidemic, the port of Callao and Lima were devastated. The first case of yellow fever was documented in Callao in February, and by the time the scourge had run its course in July, over 10 percent of Lima's population of 100,000 had died of yellow fever, Jorge Basadre, *Historia de la República del Perú*, 5th ed. (Lima, 1963), 4: 1716-17.

37. NA, T600, Winslow to SecState, Paita, 8 December 1863, N 15.

38. Holton, *Winslow Memorial*, 2: 810. A scientist to the end, in his will Winslow left specific instructions for the removal and embalming of his heart. He also specified how it was to be prepared for shipping to Massachusetts, where it was to be interred between the graves of his mother and father on Nantucket Island. Winslow instructed his executor to have his body cremated "in a retort, as in now practiced in such cases in Germany," and to ship the ashes to be interred on top of his wife's coffin at Mount Auburn Cemetery near Boston, Holton, *Winslow Memorial*, 2: 813.

Chapter 10

1. Louis Clinton Nolan, "The Relations of the United States and Peru with Respect to Claims, 1822-1870," *Hispanic American Historical Review* 17:1 (February 1937): 40.

2. On the Castilla government, see Jorge Basadre, *Historia de la República del Perú*, 5th ed. (Lima, 1963), 2: 727-851, 3: 1097.

3. W. M. Mathew, *The House of Gibbs and the Peruvian Guano Monopoly* (London, 1981), 2.

4. National Archives of the United States, Diplomatic and Consular Records, RG 59 (hereafter cited as NA), T600, Ringgold to SecState, Paita, 10 January 1854, N 6. For comparison, in 1866 a French traveler who visited Trujillo, the capital of the Department of La Libertad, estimated the population at only 6,000, of which only about 5 percent were, in his words, "Europeans." Trujillo did boast, however, a Chinese restaurant named *La Bola de Oro*, Alexis, Compte de Gabriac, *Promenade à travers l'Amérique su sud, Nouvelle-Grenade, Equateur, Pérou, Brésil;...* (Paris, 1866), 170.

5. Henry Willis Baxley, *What I Saw on the West Coast of South*

and North America, and at the Hawaiian Islands (New York, 1865), 61; Madeleine Vinton Dahlgren, *South Sea Sketches; a Narrative* (Boston, 1881), 233; George E. Squier, *Peru: Incidents of Travel and Exploration in the Land of the Incas* (New York, 1877), 26.

6. Charles M. Pepper, *Panama to Patagonia: the Isthmian Canal and the West Coast Countries of South America* (Chicago, 1906), 75, 78

7. Baxley, *What I Saw on the West Coast*, 64-65.

8. [International] Bureau of the American Republics, *Peru, Bulletin No 60* (Washington, D.C., revised to 1 May 1895), 97.

9. Jorge Basadre, *Historia de la República del Perú*, 5th ed. (Lima, 1963) 3:1252-53; Alexander George Findlay, *A Directory for the Navigation of the South Pacific Ocean; with Descriptions of its Coasts, Islands, etc. from the Strait of Magalhaens to Panama,...*, 4th ed. (London, 1877), 241; *Peru, Bulletin No. 60*, 80-81.

10. NA, T600, Ringgold to SecState, Paita, 12 April 1854; Basadre, *Historia de la República del Perú*, 3: 1252-53; Baxley, *What I Saw on the West Coast*, 61; *Peru, Bulletin No. 60*, 96.

11. Gabriac, *Promenade à travers l'Amérique su Sud*, 142; Baxley, *What I Saw on the West Coast*, 61; Squier, *Peru: Incidents of Travel*, 26. Gabriac states that the prefabricated iron building in Paita was made in the United States, but that is not likely.

12. NA, T600, Ruden to SecState, Paita, 14 July 1846, N 55; *Peru, Bulletin No. 60*, 41. To put into perspective this reference to the installation of a provincial telephone network in Northern Peru in the late nineteenth century, I note that in 1887, a decade or so after Alexander Graham Bell was granted a United States patent for a device that would transmit speech sounds over electric wires, there were more than 150,000 telephones in the United States, but only 9,000 in France and even fewer in Russia, *Encyclopedia Brittanica*, 2: 615.

13. NA, T600, Winslow to SecState, Paita, 8 December 1863, N 15; Basadre, *Historia de la República del Perú*, 2: 951; William S. Bell, *An Essay on the Peruvian Cotton Industry, 1825-1920* (Liverpool, 1985), 32.

14. H. J. Mozans [pseud. for John Augustine Zahm], *Along the Andes and Down the Amazon* (New York and London, 1911), 116-17.

15. NA, T600, Ringgold to SecState, Paita, 10 January 1854, N 6.

16. C. Reginald Enock, *The Andes and the Amazon: Life and Travel in Peru* (London, 1907), 164; Pepper, *Panama to Patagonia*, 147,

gives the altitude of this pass as 7,200 feet.

17. Enock, *The Andes and the Amazon*, 164; Mozans, *Along the Andes*, 117.

18. Findlay, *A Directory for the Navigation*, 241; Mariano Felipe Paz Soldan, *Diccionario geográfico estadístico del Perú...* (Lima, 1877), 760; *Peru, Bulletin No. 60*, 82, 33-34, 42, 96; Alejandro Garland, *Peru in 1906, with a Brief Historical and Geographical Sketch* (Lima, 1907), 251; Pepper, *Panama to Patagonia*, 78; NA, T600, Murphy to Second Assistant SecState, Paita, 30 October 1871.

19. NA, T600, Murphy to Second Assistant SecState, Paita, 30 October 1871; NA, T393, *1871 - Neuvo itinerario de los vapores entre Panamá, Guayaquil, Payta, Callao, Valparaiso e intermedios*, Callao, 22 March 1871.

20. NA, T600, Ruden to SecState, Paita, 1 January 1845; Ringgold to SecState, Arica, 1 March 1850, N 2; Paita, 10 January 1854, N 6; 12 April 1854; Gabriac, *Promenade à travers l'Amérique su Sud*, 143, 120; *Peru, Bulletin No. 60*, 38-39; NA, T393, *Gran rebaja en las tarifas de pasajes. Compañía de Navegación por Vapor en el Pacífico. Tarifa de pasajes por los vapores intermediarios entre el Callao y Tumbes*, 1 March 1869; Murphy to Second Assistant SecState, Paita, 30 October 1871.

21. *Peru, Bulletin No. 60*, 97, 81; Pepper, *Panama Patagonia*, 78.

22. *Peru, Bulletin No. 60*, 97.

23. NA, T600, Murphy to Second Assistant SecState, Paita, 30 October 1871.

24. Rory Miller, "Small Business in the Peruvian Oil Industry: Lobitos Oilfields Limited Before 1934," *Business History Review* 56:3 (1982): 400-01; Pepper, *Panama to Patagonia*, 78.

25. *Peru, Bulletin No. 60*, 97; Pepper, *Panama to Patagonia*, 75.

26. Basadre, *Historia de la República del Perú*, 2: 667, 650; Berthold Carl Seeman, *Narrative of the Voyage of H.M.S. Herald During the Years, 1845-51, under the command of Captain Henry Kellett, R.N., C.B.; Being a Circumnavigation of the Globe...*, 2 vols. (London, 1853), 152-53.

27. *Peru, Bulletin No. 60*, 98; Murphy to Second Assistant SecState, Paita, 31 March 1872.

Chapter 11

1. Madeleine Vinton Dahlgren, *South Sea Sketches; A Narrative* (Boston, 1881), 234.

2. National Archives of the United States, Diplomatic and Consular Records, RG 59 (hereafter cited as NA), T600, Murphy to SecState, New York, 25 January 1871; 1 February 1870 [*sic*, 71]; 14 February 1871; 3 March 1871; Murphy to Second Assistant SecState, Paita, 24 April 1871; NA, M968, John Fitch to Hamilton Fish, New York, 1 November 1870; Saratoga Springs, 9 July 1875; Petition from Murphy to President Grant, endorsed by Fitch and others, [February 1871]. On the battle of Cold Harbor, see *Encyclopedia Americana*, 1990, 7: 221.

3. NA, M968, Petition from Murphy to President Grant, endorsed by Fitch and others, [February 1871]. NA, T600, Murphy to Second Assistant SecState, Paita, 14 December 1874. After leaving Paita, Murphy was assigned to the United States consulate at Minatitlán, a river port in the State of Vera Cruz, Mexico. After that consulate was closed, Murphy returned to Brooklyn, but a recurrance of his tuberculosis impelled him in October of 1875 to again seek a consular assignment. NA, T600, Murphy to Hamilton Fish, Brooklyn, 30 October 1875.

4. NA, T600, Columbus to SecState, Paita, 30 September 1868, N 4.

5. NA, T600, Murphy to Second Assistant SecState, Paita, 30 April 1873.

6. NA, T600, Murphy to Second Assistant SecState, Paita, 18 June 1874, N 20.

7. All of the data on the murder of Leipsker is taken from NA, T600, Columbus to SecState, Paita, 16 December 1870, N9; 20 February 1871; Murphy to Second Assistant SecState, Paita, 24 April 1871; 24 December 1871; 31 March 1872; and an undated clipping from the Panama City *Star and Herald*.

8. Consul Murphy, in describing the *pletina* used on Major Varea in Piura, may have helped to resolve an etymological enigma. The dictionary of the *Real Academia Española* defines *pletina* as a long metal bar that is thicker than it is wide, J. Corominas, *Diccionario crítico etimológico de la lengua castellana* (Madrid, 1954), 3: 822, cites this definition, and adds that the Academy "*no nos informa ni del empleo que se da ni del oficio a que pertenece*" ("it does not reveal how it is used or the activity to which it belongs").

9. José Balta, who owned and farmed land in the Chiclayo area in the 1850s, was a key player in the success of the 1867 revolution of General Pedro Diez Canseco against the Prado government, mentioned in chapter 3. Balta's military takeover of Chiclayo in late 1867 and successful resistance to a 26-day seige laid by troops loyal

to the Prado government are documented in a fascinating 13-page report by the United States consul in Lambayeque, see NA, T393, Montjoy to SecState, Lambayeque, 14 February 1868 and subsequent despatches. Balta was inaugurated president in August of 1868, as a result of elections called earlier in the year by his mentor, interim-president Diez Canseco. On the July 1872 revolt against President Balta and his ruthless henchmen, the four Gutiérrez brothers, which resulted in riots of a magnitude and savagery that had rarely been witnessed in Lima, see Jorge Basadre, *Historia de la República del Perú*, 5th ed., (Lima, 1963), 4: 1932-39.

Chapter 12

1. According to Fischer and Dornbush, alternating inflation and deflation during the period 1860-1905 resulted in exactly the same price levels for the two years. Since 1900, however, prices have increased more than 17 times, which accounts for the impressive contemporary value of the oil cargoes reported at Paita in 1857, Stanley Fischer and Rudiger Dornbush, *Economics*, 2nd ed. (New York, 1988), 613.

2. For a more complete discussion of United States medical education and the professional job market, see William G. Rothstein, *American Physicians in the Nineteenth Century; From Sects to Science* (Baltimore and London, 1972), 85-100, 108-14.

3. H. J. Mozans [pseud. for John Augustine Zahm], *Along the Andes and Down the Amazon* (New York and London, 1911), 115.

Glossary of Spanish and Hispanic-American Terms Used in the Text

Aduana - Customhouse, or provincial office of internal transit taxes on domestic trade.

Algarrobo - Carob, a leguminous tree bearing long, flat, leathery brown pods with a sweet pulp.

Arroba - A measure of weight equivalent to 25 pounds.

Bajareque - In Spanish America, a wall built of bamboo plastered with mud; a house built using the same materials and technique.

Balsa - A raft made by lashing together logs, particularly those of the balsa tree (*Ochroma lagopus*).

Bayeta - Baize, a thick flannel.

Cantina - Bar, public house.

GLOSSARY

Cascarilla - Cinchona, Jesuit's bark, Peruvian bark.

Chacra - Farm, garden plot.

Cholo - Mixed-blood, offspring of European and Amerind.

Encomienda - In sixteenth-century Spanish America, a grant to a European colonist by the crown of a community of Indians, giving the grantee the right to use their services and obliging him to oversee their conversion to Christianity.

Fonda - Inn, boardinghouse.

Jarana - A popular dance in Peru and Bolivia.

Mestizo - Crossbreed, the offspring of persons of different races, especially the child of a European and an Amerind.

Obraje - Workshop, primitive textile manufactory.

Pulperia - In Spanish America, a store where general merchandise, liquor, and food is sold.

Quintal - Hundredweight.

Real - In colonial Spanish America, the eighth part of a peso.

Sol (es) - Unit of Peruvian currency of fluctuating value.

Tablazo - In Peru, the flat desert plain characteristic of the northern Pacific coast.

Tambo - In Andean America, a combination post house, inn, and warehouse for storing goods.

Tocuyo - A coarse cotton cloth.

Vara - A Spanish lineal measure equivalent to .84 meters.

Vegas - Well-irrigated bottom land in river valleys.

Zambo - Offspring of Amerind and Black.

216

Bibliography

Archival Sources

National Archives, Washington, D.C.

Record Group 59, General Records of the Department of State (microfilm)

> T600, Dispatches from United States Consuls in Paita, 1833-74.
>
> T353, Dispatches from United States Consuls in Tumbes 1852-74.
>
> T393, Dispatches from United States Consuls in Lambayeque, 1860-88.

Letters of Application and Recommendation During the Administrations of:

> Andrew Jackson, 1829-37, M639.
>
> Martin Van Buren, William Henry Harrison, and John Tyler, 1837-45, M687.
>
> James Polk, Zachary Taylor, and Millard Fillmore, 1845-53, M873.
>
> Franklin Pierce and James Buchanan, 1853-61, M967.
>
> Abraham Lincoln and Andrew Johnson, 1861-69, M650.
>
> Ulysses S. Grant, 1869-77, M968.

New Bedford Whaling Museum, New Bedford, Massachusetts ,Manuscript Collections:

> Swift and Allen, No. 5.
>
> John Russel Thornton, No. 16.
>
> Jonathan Bourne, Jr., No. 18.
>
> J. and W. R. Wing, No. 35.
>
> Charles W. Morgan, No. 41.

BIBLIOGRAPHY

Charles Cummings, No. 51.
Thomas K. Knowles, No. 55.
Wood and Nye, No. 66.
Swift and Allen, No. 78.
Cory, No. 80.
Andrew Hicks, No. 105.

Kendall Whaling Museum, Sharon, Massachusetts, Manuscript Collection.

Mystic Seaport Museum, Mystic, Connecticut. G.W. Blunt White Library, Manuscripts Collection.
Osborn, James C. Journal, ship *Charles W. Morgan*
of New Bedford, 1841-43, Log 143.
Journal, ship *Chelsea* of New London, 1831-34, Log
371.
Journal, ship *Mercury* of New Bedford, 1837-40, Log
67
Boyd, Thomas. Journal, U.S.S. *Brandywine*, 1826-
28, Log 628
Roe, Thomas. Journal, ship *Chelsea* of New London,
1831-32, Microfilm 14.

The Church of Jesus Christ of Latter Day Saints, Salt Lake City, Utah.
Family History Library (microfilm)
Iglesia Catolica, San Francisco (Paita) Registros Parroquiales,
1835-1932.
Indice de bautismos (Paita), 1880-99.
Bautismos (Paita), 1880-1901.
Bautismos (Colan), 1835-1900.

Printed Sources

Aldana, Susana. *Empresas coloniales: las tinas de jabón en Piura*. Lima,
n. d.
Almanach de Gotha. *Annuaire diplomatique et statistique pour l'annee...*
Gotha, 1830-65.
Assadourian, Carlos Sempat. *El sistema de la economia colonial: mercado
interno, regiones y espacio económico*. Lima, 1982.
Basadre, Jorge. *Historia de la República del Perú*, 5th ed. Lima, 1963.

BIBLIOGRAPHY

Basadre, Jorge. *Introduccion a las bases documentales para a historia de la República del Perú; con algunas reflexiones*, 2 vols. Lima, 1971.

Baxley, Henry Willis. *What I Saw on the West Coast of South and North America, and at the Hawaiian Islands*. New York, 1865.

Beale, Thomas. *The Natural History of the Sperm Whale;...* London, 1839.

Bell, William S. *An Essay on the Peruvian Cotton Industry, 1825-1920*. Liverpool, 1985.

Bourne, Russell. *The View From Front Street: Travels Through New England's Historic Fishing Communities*. New York and London, 1989.

Boyd's Washington and Georgetown Directory for 1858.

Brown, Kendall W. *Bourbons and Brandy: Imperial Reform in Eighteenth-Century Arequipa*. Albuquerque, 1986.

Bueno, Cosme. *Geografía del Perú virreinal (siglo XVIII)*, publicado por Daniel Valcarcel. Lima, 1951.

Bushnell, David and Neill Macaulay. *The Emergence of Latin America in the Nineteenth Century*. New York, 1988.

Catalogue of the Trustees, Officers, & Students of the University of Pennsylvania, Session 1846-47. Philadelphia, 1847.

Clayton, Lawrence A. *Grace: W.R. Grace & Co., the Formative Years, 1850-1930*. Ottawa, Illinois, 1985.

Clayton, Lawrence A. *Los astilleros de Guayaquil colonial*, Guayaquil, 1978.

Collin Delavaud, Claude. *Les régions côtiéres du Pérou septentrional; occupation du sol, amenagement regional*. Lima, 1968.

Conner, Susan P. "Politics, Prostitution and the Pox in Revolutionary Paris, 1789-1799." *Journal of Social History* 22 (1989): 713-34.

Corning, Howard. "Sullivan Dorr, China Trader." *Rhode Island History* 3:3 (1944): 75-90.

Corominas, J. *Diccionario crítico etimológico de la lengue castellana*. Madrid, 1954.

Dahlgren, Madeleine Vinton. *South Sea Sketches; a Narrative*. Boston, 1881.

Dawson, Frank Griffin. *The First Latin American Debt Crisis: the City of London and the 1822-25 Loan Bubble*. New Haven, 1990.

Deustua Pimentel, Carlos. *Las intentencias en el Perú (1790-1796)*. Seville, 1965.

DeVere, Maximilian Schele. *Students of the University of Virginia*. Baltimore, 1878.

Dictionary of American Biography, edited by Allen Johnson. New York, 1929.

Doggett's The New-York City and Co-Partnership Directory.

BIBLIOGRAPHY

Drake, Francis S. *The Town of Roxbury; its Memorable Persons and Places; its History and Antiquities,...* Boston, n.d.

Druett, Joan, ed. *"She Was A Sister Sailor": The Whaling Journals of Mary Brewster, 1845-1851.* Mystic, 1992.

Duvall, Alfred. *Communication in Relation to a Supply of Water for the City of Baltimore, from the Gunpowder River.* Baltimore, 1854.

Enock, C. Reginald. *The Andes and the Amazon: Life and Travel in Peru.* London, 1907.

Essex Institute. *Vital Records of Roxbury Massachusetts to the end of the Year 1849.* 2 vols. Salem, 1925.

Farr, James. "A Slow Boat to Nowhere: the Multi-Racial Crews of the American Whaling Industry." *Journal of Negro History*, 68:2 (1983): 159-70.

Findlay, Alexander George. *A Directory for the Navigation of the South Pacific Ocean; with Descriptions of its Coasts, Islands, etc. from the Strait of Magalhaens to Panama,...*4th ed. London, 1877.

First Two Centuries of the Washington County Courthouse. n.p., n.d.

Fisher, John Robert. *Government and Society in Colonial Peru: the Intendent System, 1784-1814.* London, 1970.

Funnell, William. *A Voyage Round the World. Containing an Account of Captain Dampier's Expedition into the South-Seas in the Ship St. George in the Years 1703 and 1704.* 1707. Reprint, Amsterdam, 1969.

Gabriac, Alexis, Compte de. *Promenade à travers l'Amérique su Sud, Nouvelle-Grenade, Equateur, Pérou, Brésil;...* Paris, 1868.

Garland, Alejandro. *Peru in 1906, with a Brief Historical And Geographical Sketch.* Lima, 1907.

*Genealogy and Biography of Leading Families of the City of Baltimore and Baltimore County, Maryland...*New York, 1897.

Gootenberg, Paul. *Between Silver and Guano: Commercial Policy and the State in Postindependence Peru.* Princeton, 1989.

Guy, Donna J. "White Slavery, Public Health, and the Socialist Position on Legalized Prostitution in Argentina, 1913-1936." *Latin American Research Review* 23:3 (1988): 60-80.

Haenke, Tadeo. *Descripción del Perú.* Lima, 1901.

Hamlin, Talbot. *Benjamin Henry Latrobe.* New York, 1955.

Hohman, Elmo Paul. *The American Whaleman: A Study of Life and Labor in the Whaling Industry.* 1928. Reprint, Clifton, New Jersey, 1972.

Holton, David-Parson. *Winslow Memorial. Family Records of Winslows and their Descendants in America,...*New York, 1888.

Index to Wills for New York County, 1662-1850.

[International] Bureau of the American Republics, *Peru, Bulletin No.*

220

BIBLIOGRAPHY

60, 1892. Washington, revised to 1 May 1895.

Kubler, George. *The Indian Caste of Peru, 1795-1940. A Population Study Based upon Tax Records and Census Reports*. Washington, D.C., 1952.

Lafond [de Lurcy], G[abriel]. *Voyages autour du monde et naufrages célebres*. 2 vols. Paris, 1843.

Lane, (Mrs.) Julian C. [Janie W.]. *Key and Allied Families*. Macon, Georgia, 1931.

Langdon, Robert, *American Whalers and Traders in the Pacific: A Guide to Records on Microfilm*. Canberra, 1978.

Langdon, Robert. *Thar She Went: An Interim Index to Pacific Ports and Islands Visited by American Whalers and Traders in the 19th Century*. Canberra, 1979.

Lessón, René-Primevère. *Voyage autour du monde entrepris par ordre du Gouvernement sur la corvette* La Coquille. 1838, reproduced in Comisión Nacional del Sesquicentenario de la Independencia del Perú, *Colección documental del la independencia del Perú. tomo XXVII Relaciones de viajeros*, 2 vols. Lima, 1971.

Lofstrom, William. *Dámaso de Uriburu: un empresario minero de principios del siglo XIX en Bolivia*. La Paz, 1982.

Lofstrom, William. *La Presidencia de Sucre en Bolivia*. Caracas, 1987.

Longworth's New York Directory.

Lorente, Sebastián. *Historia de la consquista del Perú*. Lima, 1861.

Mathew, W.M. *The House of Gibbs and the Peruvian Guano Monopoly*. London, 1981.

McCreery, David. "'This Life of Misery and Shame:' Female Prostitution in Guatemala City, 1880-1920." *Journal of Latin American Studies*. 18 (1986): 333-53.

McIntire, Robert Harry. *Annapolis Maryland Families*. Baltimore, 1979.

Meyer, Michael C., and William L. Sherman. *The Course of Mexican History*. 4th ed. New York, 1990.

Miller, John. *Memoirs of General [William] Miller, in the Service of the Republic of Peru*. 2 vols. London, 1928.

Miller, Rory. "Small Business in the Peruvian Oil Industry: Lobitos Oilfields Limited Before 1934." *Business History Review* 56:3 (1982): 400-01.

Mozans, H.J. [pseud. for John Augustine Zahm]. *Along the Andes and Down the Amazon*. New York and London, 1911.

National Archives Trust Fund Board, National Archives and Record Administration. *Diplomatic Records; A Select Catalog of National Archives Microfilm Publications*. Washington, D.C., 1986.

Newman, Harry Wright. *Mareen Duvall of Middle Plantation; A Gen-*

221

BIBLIOGRAPHY

*ealogical History of Mareen Duvall,...*Washington, D.C., 1952.

Niles Weekly Register. Baltimore.

Nolan, Louis Clinton. "The Relations of the United States and Peru with Respect to Claims, 1822-1870." *Hispanic American Historical Review* 17:1 (February 1937): 40.

Pareja Diezcanseco, Alfredo. *Historia del Ecuador.* Quito, 1961.

Paz Soldan, Mariano Felipe. *Diccionario geográfico estadístico del Perú...*Lima, 1877.

Pease, Zephaniah W. *History of New Bedford.* 2 vols. New York, 1918.

Pepper, Charles M. *Panama to Patagonia: the Isthmian Canal and the West Coast Countries of South America.* Chicago, 1906.

Prescott, William H. *History of the Conquest of Peru, with a Preliminary View of the Civilization of the Incas.* 2 vols. New York, 1851.

Proctor, Robert. *Narrative of a Journey Across the Cordillera of the Andes, and of a Residence in Lima, and Other Parts of Peru, in the Years 1823 and 1824.* London, 1825.

Raimondi, Antonio. *El Perú: tomo II, historia de la geografía del Perú.* Lima, 1876.

A Report of the Record Commissioners of the City of Boston, Containing Boston Births from A.D. 1700 to A.D. 1800. Boston, 1894.

Richardson III, James B., and Elena B. Decima Zamecnik. "The Economic Impact of Martha's Vineyard Whalers on the Peruvian Port of Paita." *The Dukes County Intelligencer* 10:3 (February 1977): 67-91.

Ricketson, Daniel. *The History of New Bedford, Bristol County, Massachusetts:...*New Bedford, 1858.

Rothstein, William G. *American Physicians in the Nineteenth Century; from Sects to Science.* Baltimore and London, 1972.

Ruggles, Steven. "Fallen Women: the Inmates of the Magdalen Society Asylum of Philadelphia, 1836-1908." *Journal of Social History* 16 (Summer 1983): 65-82.

Rumazo González, Alfonso. *Manuela Sáenz, la libertadora del Libertador.* Bogota, [1945].

[Ruschenberger, William S.W.] *Three Years in the Pacific; Including Notices of Brazil, Chile, Bolivia and Peru.* Philadelphia, 1834.

Sailors Magazine and Naval Journal. 1836.

Salle, A. de la. *Voyage autour du monde éxecuté pendant les années 1836 et 1837 sur la corvette* La Bonite,... *relation du voyage,* 3 vols. Paris, 1851.

Santa-Cruz Schuhdrafft, Andres de. *Cuadros sinopticos de los gobernantes de la Republica de Bolivia, 1825 a 1956, y de la del Peru, 1820-1956.* La Paz, 1956.

BIBLIOGRAPHY

Schoch, Mildred Cook. *Ringgold in the United States*. privately printed, 1970.

Seeman, Berthold Carl. *Narrative of the Voyage of H.M.S. Herald During the Years 1845-51, under the Command of Captain Henry Kellett, R.N., C.B.; Being a Circumnavigation of the Globe,*....2 vols. London, 1853.

Sherman, Stuart C. *Whaling Logbooks and Journals, 1613-1927: An Inventory of Manuscript Records in Public Collections*. New York, 1986.

Smith II, Walter B. *America's Diplomats and Consuls of 1776-1865*. U.S. Department of State, Foreign Service Institute, Center for the Study of Foreign Affairs, Occasional Paper N 2. Washington, D.C., 1986.

Spindler, Robert P. "Synopsis of the Dorr Family Papers," Phillips Library, Peabody & Essex Museum, Salem, Massachusetts, May 1988.

Spruce, Richard. *Notes on the Valleys of Piura and Chira, in Northern Peru, and on the Cultivation of Cotton Therein*. London, 1864.

Squier, E. George. *Peru: Incidents of Travel and Exploration in the Land of the Incas*. New York, 1877.

Stackpole, Edouard A. *The Sea-Hunters; The New England Whalemen During Two Centuries, 1635-1835*. Philadelphia and New York, 1953.

Stackpole, Edouard A. *Whales & Destiny; The Rivalry Between America, France, and Britain for Control of the Southern Fishery, 1785-1825*. Amherst, Massachusetts, 1972.

Starbuck, Alexander. *History of the American Whale Fishery from its Earliest Inception to the Year 1876*. 1878. Reprint, New York, 1964.

Stevenson, William Bennet. *A Historical and Descriptive Narrative of Twenty Years' Residence in South America, in Three Volumes*. London, 1825.

Stewart, Watt. *Chinese Bondage in Peru: A History of the Chinese Coolies in Peru, 1849-1874*. Durham, North Carolina, 1951.

Stiglich, German. *Diccionario geográfico del Perú*. Lima, 1922.

The Concise Dictionary of National Biography, part I. London, 1953.

The Roxbury [Massachusetts] Directory: 1852. Roxbury, 1852.

The Washington Directory, and Congressional and Executive Register, for 1850. Compiled by Edward Waite. Washington, 1850.

The Washington Directory, and Governmental Register, for 1843;... Compiled by Anthony Reintzel. Washington, D.C. 1843.

The Washington Directory, and National Register, for 1846,... Washington, 1846.

The Washington and Georgetown Directory, Strangers' guide Book for Washington,...for 1853.

Thwing, Walter Eliot. *History of the First Church in Roxbury, Massachusetts, 1860-1904*. Boston, 1908.

BIBLIOGRAPHY

Torrey, William. *Torrey's Narrative: or, the Life and Adventures of William Torrey*...Boston, 1848.

University of Virginia Catalogue, 1846-47.

Walkers' Buffalo City Directory, n.d.

Walker, Homer A. *Historical Court Records of Washington, District of Columbia*, vols. 8-11. Marriages Joy, T. - Rophinot. n.p., n.d.

Walters, Richard. *A Voyage Round the World, in the Years MDCCXL, I, II, III, IV. by George Anson, Esq; Commander in Chief of a Squadron of His Majesty's Ships, Sent Upon an Expedition to the South Seas*... London, 1748.

Washington County, Maryland, Church Records of the 18th Century. Westminster, Maryland. Family Line Publications, 1988.

Williams, Thomas J.C. *A History of Washington County, Maryland, from the Earliest Settlements to the Present Time*....2 vols. n.p., 1906.

Winslow, Charles F. "Letters from California." *The Magazine of History* 103 (1924).

Index

INDEX

INDEX

INDEX

INDEX